ZEOLITES - USEFUL MINERALS

Edited by **Claudia Belviso**

Zeolites - Useful Minerals
http://dx.doi.org/10.5772/61950
Edited by Claudia Belviso

Contributors

Kristin Gleichmann, Baldur Unger, Alfons Brandt, T.V. Ojumu, James Brassell, Aleksadra Badora, Changjiu Xia, Min Lin, Bin Zhu, Xingtian Shu, YiBin Luo, Minoru Fujii, Hui Lin, Daniel G Bucur, Maria Harja, Pedro Augusto Arroyo, Lidiane Sabino Da Silva, Victor Silva, Maria Angelica Barros, Jiwan Singh, Ajay S. Kalamdhad, Byeong-Kyu Lee, Takafumi Miyanaga, Yushi Suzuki, Sho Narita, Magdalena Wdowin, Mugera Gitari, Leslie Petrik, Nicholas Musyoka

Published by InTech

Janeza Trdine 9, 51000 Rijeka, Croatia

Publishing Process Manager Edi Lipovic

Technical Editor SPi Global

Cover InTech Design team

First published August, 2016

Printed in Croatia

Additional hard copies can be obtained from orders@intechopen.com

Zeolites - Useful Minerals, Edited by Claudia Belviso

p. cm.

Print ISBN 978-953-51-2576-1
Online ISBN 978-953-51-2577-8

For EU product safety concerns:
IN TECH d.o.o., Prolaz Marije Krucifikse Kozulić 3, 51000 Rijeka, Croatia,
info@intechopen.com or visit our website at intechopen.com.

Meet the editor

Claudia Belviso is a researcher at the Institute of Methodologies of Environmental Analysis (IMAA) of CNR. Since 2002 she has been carrying out her research activity in the field of mineralogy and geochemistry aimed at environmental protection. She is responsible for research activity on zeolite synthesis from waste materials as well as natural sources and their application to solve environmental problems. This activity has allowed her to be the inventor of an international patent about zeolite synthesis from fly ash, to publish numerous scientific articles in peer-reviewed journals; to participate in many national and international conferences; to take part in the organisation of international congresses and national and international schools and to carry out scientific researches in national and international projects supported by public and private institutions.

Contents

Preface

The book collects recent results about research activities on zeolites, from synthesis to application. It is composed of two sections. The first is devoted to articles and brief review articles on the synthesis of zeolite from fly ash and final application of these newly formed minerals to solve environmental problems. The second part of the book provides useful information on different applications both of natural and synthetic zeolites ranging from environmental pollution to industrial and commercial applications. The performance of zeolite molecular sieves, hollow titanium zeolites and luminescent zeolites is interesting considering the new frontiers reached by the research on zeolites.

This book is a useful instrument for readers interested in investigating innovative aspects of the studies on zeolite, and I want to thank all the authors for their significant contribution.

Zeolites are hydrated aluminosilicate minerals that can be synthesised using several source materials.

A number of water purification processes have been developed in recent years based on the utilisation of low-cost materials with high pollutant removal efficiency.

Zeolites are microporous crystalline hydrated aluminosilicates characterised by a three-dimensional network of tetrahedral (Si, Al)O4 units that form a system of interconnected pores. The aluminium ion produces a net negative charge, which is balanced by the presence of an extra cation in the framework. One important property of zeolite is its ability to exchange cations. Due to their similar structure and physicochemical properties, synthetic zeolites have replaced natural zeolites in a variety of applications.

Zeolites are hydrated aluminosilicate minerals consisting of a three-dimensional open structure with excellent ion-exchange and sorption properties, which makes them notably useful for resolving the mobility of toxic elements in a number of environmental applications. Their effectiveness is strictly connected to their typical structural characteristics, which facilitate pollutant adsorption and encapsulation.

Fly ash is a product of coal combustion waste that consists of an amorphous aluminosilicate glass, quartz and mullite. The high glass content makes fly ash a suitable material for the synthesis of zeolites, which are hydrated aluminosilicate minerals with a three-dimensional open structure. Zeolites are useful in a number of environmental applications.

Coal fly ash is the most abundant coal combustion by-product, partly used in concrete and cement manufacturing yet more than half of it disposed of in landfills since it finds no other application.

The huge production of fly ash is extremely of concern because of this kind of disposal, and several investigations have been carried out to try to exploit this waste material. Over the last few years, fly ash has been gaining ground in the synthesis of zeolites. This is due to the fact that the main fly ash component is amorphous aluminosilicate glass, which is the prevalent reactive phase.

Dr. Claudia Belviso
Institute of Methodologies for Environmental Analysis
CNR-IMAA, Italy

Section 1

Fly Ash and Zeolite

Chapter 1

Upscaling of Zeolite Synthesis from Coal Fly Ash Waste: Current Status and Future Outlook

James Philip Brassell, Tunde Victor Ojumu and
Leslie Felicia Petrik

Additional information is available at the end of the chapter

http://dx.doi.org/ 10.5772/63792

Abstract

Coal fly ash has been recognised as suitable feedstock for zeolite synthesis, and with this discovery, there has been much literature produced over the years on various synthesis processes with a view to address the disposal problems associated with fly ash wastes. The different synthesis processes can be divided into a pre-synthesis activation with an alkali solution and a hydrothermal synthesis of heating the resulting aluminosilicate gel. However, the challenge lies in up scaling the synthesis to allow production on a commercial scale. The most common pre-synthesis involving high-temperature fusion of fly ash with sodium hydroxide may not be feasible in an economic context, given that a large-scale furnace that would be needed. In this review, the successes recorded in the application of acoustic cavitation by using ultrasonication technique were discussed as a possible alternative to the fusion process on one hand. On the other hand, jet-loop reactor system provides hydrodynamic cavitation, which may be relatively scalable compared to ultrasonication and may offer an economic advantage, if systematically explored, over fusion process for large-scale synthesis of zeolites. It is also recommended that agitation is critical during the crystallisation process and the suitability of an impeller type that offers a minimum shearing to be explored during the hydrothermal process with respect to quality and yield of the zeolite produced. The understanding gleaned from these recommendations may be useful in designing an appropriate scale-up operation for zeolite synthesis from coal fly ash.

Keywords: fly ash, zeolites, up scale, jet loop, hydrothermal agitation

1. Introduction

It has been 30 years since the discovery by Holler and Wirsching that zeolites can be synthesised from coal fly ash [1]. It was found that the coal fly ash contained the necessary elements, namely silicon and aluminium, which are building blocks of the zeolitic structure. This has spawned substantial research into zeolite synthesis from coal fly ash with a view to create outlets for fly ash utilisation, thereby solving the worldwide problem of coal fly ash disposal. Coal fly ash is produced during the coal combustion process, and it is the most predominant coal combustion by-product [2]. The electricity generated from coal combustion accounted for 29.9% of the world's electricity supply in 2011, and this is expected to increase 46% by 2030 [3]. Therefore, there are concerns with the high volumes of coal fly ash waste being produced.

In South Africa alone, approximately 36 million tonnes are produced per annum with 95% of that being dumped in ash dumps [4]. When it is seen that South Africa only accounts for 2.4% of the world coal consumption, it becomes apparent of the scale of coal fly ash produced worldwide. Especially, concerning the large-scale world coal consumers, such as China (50.2%), United States (11.7%), and India (8%) [3], it has been reported that China produced 540 million tonnes of coal fly ash in 2011 with only 68% being reused, while about 61% of the 163.56 million tonnes of fly ash produced in India between 2012 and 2013 was reused [3]. This poses risks to the environment and surrounding populations of these ash dumps. This is due to coal fly ash having the potential to leach toxic heavy elements into the groundwater system [5] and poses a threat to person's health if inhaled [6]. Due to the problem of coal fly ash disposal, many studies investigated its use with regard to different applications, such as in agriculture, in construction, in wastewater treatment, and for the synthesis of geopolymers and zeolites as can be seen in a recent review [3]. The views of these studies were to find outlets for the utilisation of this waste, thereby addressing the disposal problems. However, these applications have not solved the problem of coal fly ash disposal, as they are not at the scale required to deal with the high volumes of coal fly ash being produced.

There is a need to investigate further into the feasibility of scaling up some of these applications. Although recent studies have been focusing on some fundamental investigations regarding scale-up development for the synthesis of zeolite from coal fly ash [7, 8], there remains a significant amount of studies to be conducted in order to realise the end goal of synthesis on a commercial scale. This book chapter aims at giving a review of the work that has been produced regarding the synthesis of zeolites from coal fly ash with a view to recommending outstanding investigations to be carried out regarding the up scale of zeolite synthesis from coal fly ash. It will justify the necessary procedures that need to be studied in the future to realise the end goal of large-scale zeolite synthesis from coal fly ash.

2. Coal fly ash

Coal fly ash is mainly composed of silica (SiO_2), alumina (Al_2O_3), ferrous oxide (Fe_2O_3) and calcium oxide (CaO) with also some unburned carbon detected through loss on ignition tests

(LOI) [9, 10]. The chemical properties depend on the type of coal being burned, along with handling and storage procedures [3, 9, 10]. This gives rise to the high variability in coal fly ash compositions obtained from different batches of coal processed in the same power station [8]. The major and minor elements in coal fly ash are expressed as oxides, with the main ones being already mentioned previously. However, it also consists of trace elements, with some being As, B, Ba, Be, Cd, Co, Cr, Cu, Ge, Hg, Li, Mo, Ni, Pb, Rb, Sb, Se, Sn, Sr, Th, U, V and Zn [10]. The major phases of coal fly ash are amorphous glass, mullite and quartz with various others in smaller amounts [10, 11]. A recent comprehensive study by Musyoka [12] on the mineral phases of South African coal fly ash reported that the ash yielded percentage amounts ranging from the more amorphous content of 57.54% to less amorphous content of 41.55%. The mullite ranged from 34.55 to 22.27% and the quartz from 21.38 to 11.92%. Another comprehensive study was performed by Vassilev and Vassileva [11] on coal fly ash from Spain, Bulgaria, the Netherlands, Italy, Turkey and Greece. The amorphous content ranged from 85 to 34%, the mullite from 40 to 0.3% and the quartz from 14 to 2%.

Furthermore, the combustion rate and temperature of the coal combustion process affects the morphology of the coal fly ash particles [10, 13]. Coal fly ash mainly consists of spherical particles, such as solid spheres and hollow spheres (cenospheres), with some irregular unburned carbon, as determined by scanning electron microscopy (SEM) analysis [3, 13]. The colour of coal fly ash ranges from water white to yellow, orange to deep red or brown to opaque and is due to the iron and unburned carbon content [3, 14]. Coal fly ash containing more than 70% SiO_2+Al_2O_3+Fe_2O_3 and having a low lime content of 5% is classified as class F fly ash, while if the SiO_2+Al_2O_3+Fe_2O_3 content is between 50 and 70% and a high lime content of between 10 and 35%, it is classified as class C fly ash [9, 10]. It follows from the aforementioned that class C fly ash can be produced from lignite and sub-bituminous coal while class F from bituminous and anthracite coal [10, 11].

There have been numerous studies investigating the leachability of coal fly ash into the groundwater system that surrounds the ash dumpsites [5, 15–26]. In a more recent article by Nyale et al. [5], it was reported that the disposed ash at the dump site was able to release trace elements into the environment over time, and the study revealed that all the trace elements studied (As, Zn, Pb, Ni, Mo, Cr and Cu) were significantly leached from the labile phases (4.42–27.43%). The inhomogeneity of the ash dump was also attributed to the observed differences in the elements leached from varying core samples. Besides this type of environmental problem, the potential threat of coal fly ash to the health of the surrounding population and the mine workers exposed to the ash has been reported [6]. The leachability of genotoxic compounds from fly ash into human lungs if the ash dust is inhaled has been reported [6], and conditions such as lung cancer, asthma and many others are typical diseases attributed to fly ash inhalation [6, 27].

However, there have been several studies that investigated potential application of fly ash with a view to mitigate the disposal of this fly ash; some of the applications investigated include uses in agriculture, construction, extraction of rare earth elements, wastewater treatment and synthesis of geopolymers and zeolites [3]. Furthermore, only 5% of coal fly ash is being re-used as reported by Eskom [4]. A comprehensive and recent review of some of the applications of coal fly ash is reported by Yao et al. [3]; however, these authors recommended the need to explore commercialisation of the applications. The following section will focus on the various synthesis methods employed in the application of coal fly ash for zeolites synthesis and assess studies aimed towards the up scale of the synthesis process.

3. Zeolites

Zeolites are crystalline aluminosilicates of which possess cavities and pores on a molecular scale [28]. They do occur as natural minerals but are more useful being synthetically produced in laboratories for uses such as sorbents, catalysts and exchange materials [28]. The zeolite consists of SiO_4 and AlO_4 tetrahedra of which gives it an anionic framework with the negative charge of Al being compensated by extra framework cations, some being Na^+, K^+, Ca^{2+} and Mg^{2+} [29]. This can be illustrated in **Figure 1**.

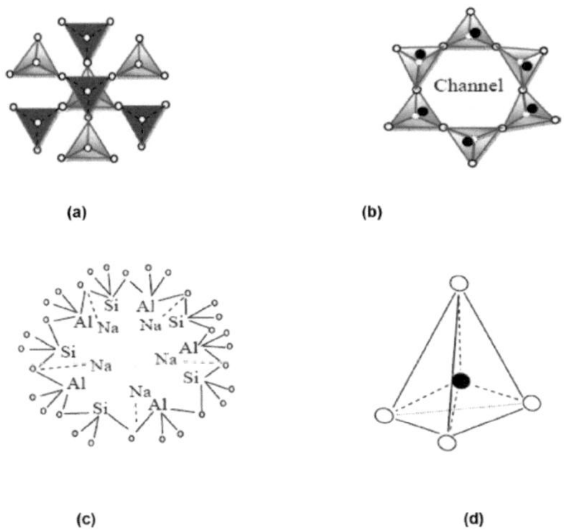

Figure 1. (a) Basic tectosilicate structure of zeolite where dark (i.e. vertex in) and light (i.e. vertex out) shades to add three-dimensional (3D) effect and upside down orientation of the tetrahedra for vertex sharing between two rings of the zeolite structure in its 2D view on a picture plane; (b) single-ring tetrahedron structure and framework of a zeolitic mineral; (c) <i>SiO_4^-</i> and <i>AlO_4^-</i> in a ring of sodium zeolite; and (d) pictorial representation of a 3D view of a tetrahedra with centrally located Si or Al atoms, exhibited by dotted lines drawn to represent the portion within the body of the mineral [30].

The primary building units of the zeolite structure are shown in **Figure 1(d)**, and they are TO_4, tetrahedra of silicon and aluminium mentioned previously [29]. There is a concept of infinite component units with the zeolite structure regarded as being made of finite component units with infinite component unit-like chains or layers, these are known as the secondary building units, some illustrated in **Figure 2**. The secondary building units can consist of single or double rings with each of those consisting of four-, five-, six- or eight-linked tetrahedral, including the silicon and aluminium atoms [30].

http://dx.doi.org/ 10.5772/63792

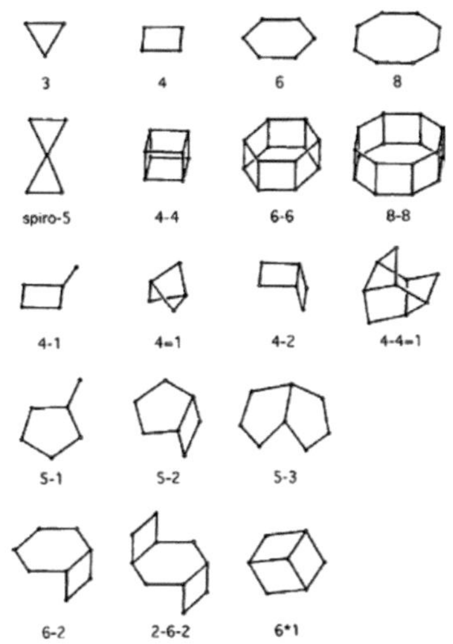

Figure 2. Secondary building units (SBU's) with SBU codes below figures [29].

The varying framework structures seen above give rise to the unique three-letter code system to name each unique zeolite, some of these codes along with the pore sizes of some zeolites are shown in **Figure 3**.

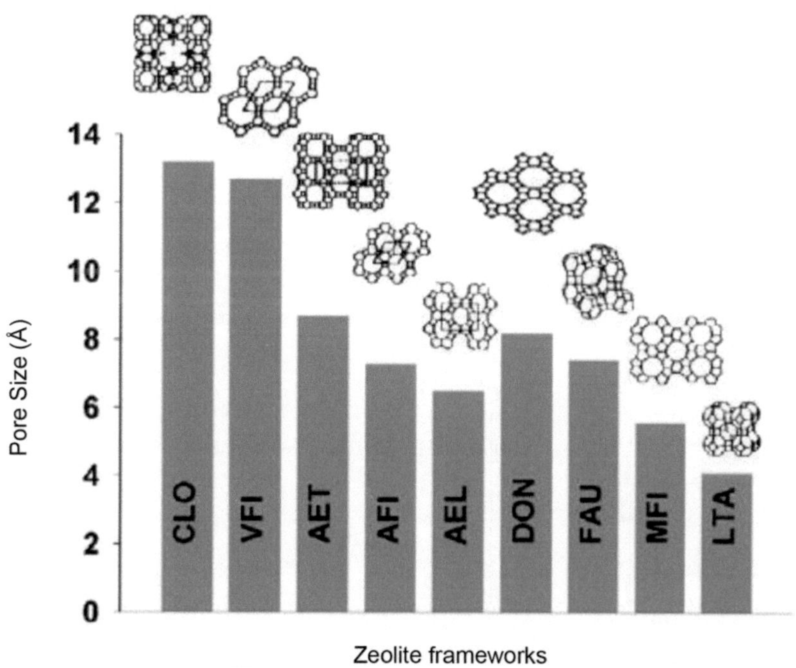

Figure 3. A comparison of the different framework pore sizes [31].

4. Synthesis of zeolites from coal fly ash

As mentioned previously, the synthesis of zeolites from coal fly ash began with the pioneering work of Holler and Wirsching in 1985 [1]. Researchers have found a number of synthesis methods to produce various types of zeolites from the coal fly ash [32–42]. All known synthetic methods can be found on the website of International Zeolite Association (IZA) at URL: http://www.iza-online.org/synthesis/default.htm [43]. However some of the prominent methods are reviewed in this section.

All the methods developed mainly involve the dissolution Al-Si bearing fly ash phases with alkaline solutions such as NaOH and KOH with the proceeding precipitation of zeolitic material [2]. The two most well-known routes for fly ash conversion to zeolite are the fusion with sodium hydroxide process developed by Shigemoto et al. [32] and a two-step process developed by Hollman et al. [35]. The two-step process developed by Hollman et al. [35], which followed from the traditional one-step process produced more "pure" zeolites; however, the process was relatively more costly. A comparison of the one-step and two-step process can be seen in **Figures 4** and **5**, respectively. Shigemoto et al. [32] developed the fusion process where coal fly ash is fused with sodium hydroxide using a high temperature, prior to hydrothermal treatment. A ratio of CFA:NaOH of 1:1.2 was optimum with a fusion temperature of 550°C, and this converted the fly ash particles into sodium salts. These salts (silicate and aluminate) after the hydrothermal treatment favoured the formation of zeolite Na-X. The process is illustrated in **Figure 6**.

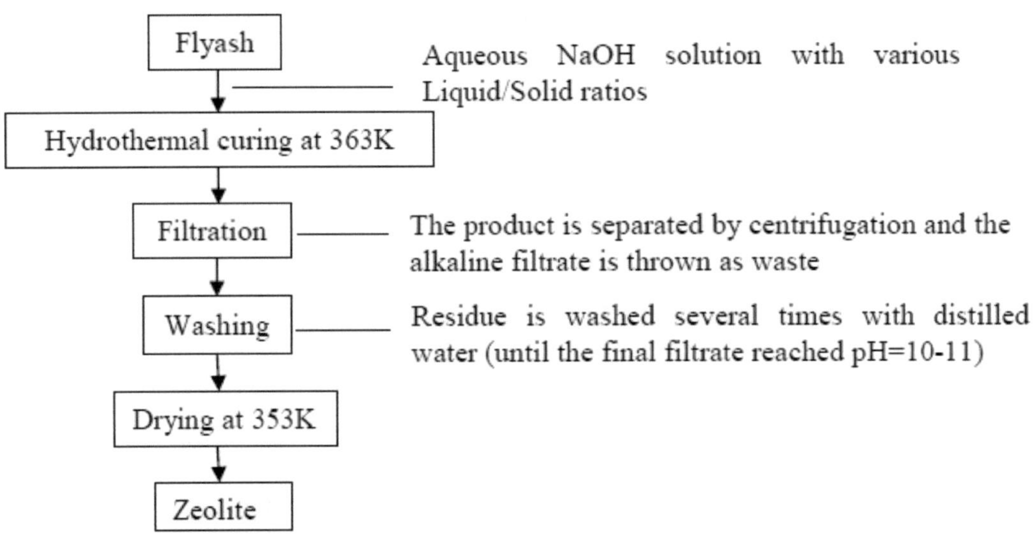

Figure 4. Flow diagram of the traditional one-step process [30].

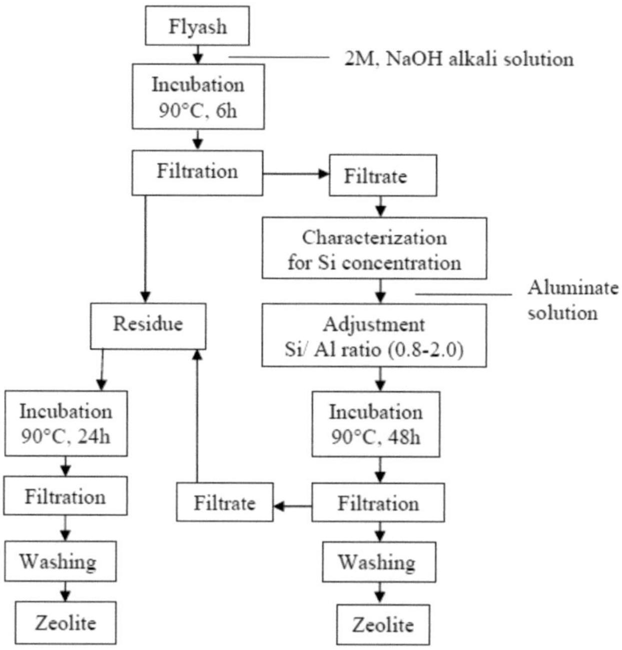

Figure 5. Flow diagram of two-step process [30].

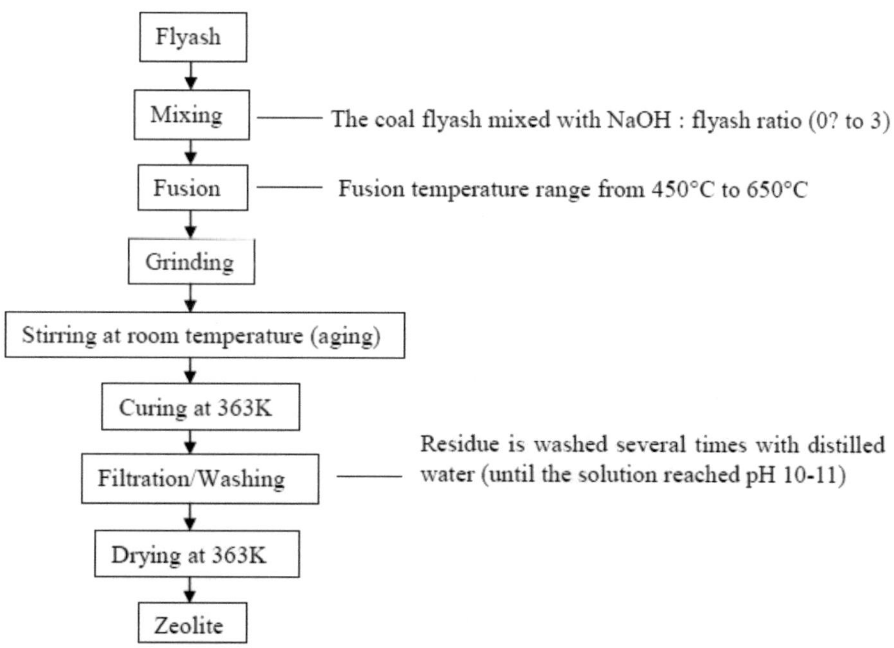

Figure 6. Flow diagram of the fusion-assisted process [30].

Murayama et al. [36] investigated the synthesis of zeolites from coal fly ash using different alkali sources, namely NaOH, Na_2CO_3 and KOH. It was found that regarding the reaction mechanism that three steps exist in an alkali hydrothermal reaction, namely dissolution, condensation and crystallisation. The dissolution step begins at a temperature range of 298–393 K with the amount of OH^- in the alkali solution making a great contribution to the reaction. The concentration of Na^+ in the alkali solution mainly determined the total reaction rate of the zeolite synthesis.

Not only the degree of zeolitisation is affected by the NaOH concentration but also the type of zeolite formed. This is according to Molina and Poole [44] who noted the difference found in the zeolite product was due to the increase in supersaturation achieved from the higher proportion of soluble species from the rise of the NaOH concentration. This phenomenon can be described by the Ostwald's rule of successive transformation with the higher the supersaturation, better the conditions are to nucleate metastable phases [44]. An example being zeolite X which later recrystallises and is replaced by the more stable zeolite hydroxysodalite. Some other factors affecting zeolite formation from coal fly ash are the SiO_2/Al_2O_3 ratio in the feedstock, the NaOH concentration as mentioned previously, the extra framework cations, the water content, the synthesis time and temperature and the agitation used in the zeolite synthesis.

Regarding the SiO_2/Al_2O_3 ratio Inada et al. [38] demonstrated the dependence of this ratio in zeolite synthesis. Two differing zeolites were formed from a silica rich fly ash (zeolite Na-P1) and an alumina-rich fly ash or silica lean fly ash (hydroxysodalite) with the ratio being controlled with SiO_2 and Al_2O_3 aerosil powders. According to Querol et al. [45], this ratio depends on the compositions of these species within the aluminosilicate glass phase of the coal fly ash and not on their bulk composition in the fly since this phase is the first to be activated.

Murayama et al. [36] also showed how the choice of cations could significantly alter the reaction to form one zeolite crystal over another. The Na^+ ions promoted the formation of zeolite P, while the K^+ ions promoted the formation of chabazite. Indeed, zeolite synthesis was shown to be favoured by the use of Na^+ ions over K^+ ions [46, 47]. This is due to Na^+ ions stabilising the secondary building units of the zeolite frameworks, while K^+ ions act as a suppressor due to their promotion of slow crystallisation rates [47].

There is a problem when it comes to zeolite synthesis regarding the water content used. This is due to the increase in water content resulting in an increase in yield when using fly ash as the starting material. This is because the higher water content results in the increase in the crystalline and amorphous phases dissolution rates in the coal fly ash [46, 48]. This led to investigation of using alternate water sources, such as investigating using tap water, distilled water and acid mine drainage (AMD) in the replacement of ultra-pure water. It was reported by Mainganye [8] that tap water and distilled water could produce similar products as the ultra-pure water, while AMD water, however, was not suitable. Musyoka et al. [49] established that AMD could yield hydroxysodalite—a zeolite with less application. However, it was found that circumneutral mine water produced a comparable quality of Na-P1 and X zeolites to ultra-pure synthesis. Zeolite X and ZK-5 were successfully synthesised using seawater at lower temperatures with hydroxysodalite also being formed; however, zeolite A could not be achieved [42]. Artificial seawater was used at room temperature (25°) to crystallise zeolites A and X in a later study [50].

The thermodynamic stability of zeolite has been reported by Boycheva et al. [51] to follow the sequence in the order of Linda --> Faujasite--> Chabazite--> Na-P1 --> Hydroxysodalite. It has been observed previously in the transformation of faujasite into more stable phases such as chabazite and Na-P1 at longer synthesis times in the reactant liquor [37, 51]. Due to this metastable behaviour of zeolite X, its formation is always competitive with hydroxysodalite, zeolite X favouring the lower temperatures [42, 51]. It is also known that a higher temperature produces zeolites of a greater particle size [44]. It was found that aluminium dissolution tends to be faster than silicon, of which the two need to be continuously dissolved to promote and feed crystal growth, the silicon dissolution increasing with temperature [44]. Therefore, the Si/Al ratio will favour the formation of zeolite X over A at higher temperatures. When the glassy content is high, the synthesis time drops, while a high content of quartz and mullite requires longer reaction times [2].

Inhomogeneous mixtures can result due to inadequate mixing which in turn has a dramatic effect on zeolite synthesis, this is pertaining to the viscous gel produced before the crystal growth step [52]. These inhomogeneous mixtures produce "pockets" of gel each with differing compositions and each acting like a "mini-reactor", producing phases based on that "mini-reactor's" composition. The effect of agitation is not well understood and this is due to mixing having to perform the following tasks during crystallisation [52]:

- Reagent dissolution
- Initial gel formation
- Maintaining a homogenous gel
- Aiding in gel structure break-up
- Maintaining a uniform temperature across the reactor
- Transferring "nutrients" to the growing crystals
- Keeping the zeolite crystals in complete suspension on completion of the reaction

The substantial change in viscosity during synthesis is also an important factor to consider. The solution can go from a viscous gel to something resembling water during the course of synthesis [52]. Marrot et al. [53] explained another important factor in agitation, that of shearing, of which can have harmful effects on zeolite synthesis. This can affect the stability of the zeolite and the purity of which can be synthesised. The highest crystallinity was produced by the Archimedes screw impeller due to the low rate of shearing, while higher shearing resulted in lower crystallinities [53]. Agitation during the ageing step of synthesis was investigated by Mainganye et al. [54]. It was shown that agitation during the hydrothermal step of synthesis can have a harmful effect on the stability and purity of the zeolite regarding the shearing produced [53, 54]. It was thought that, during the aging step, shearing may favour the zeolite formation due to it facilitating the dissolution of coal fly ash into the alkaline solution. The four-blade impeller was found to optimum for zeolite synthesis, applied during the ageing step [54]. It is noted however that regarding the up scale of zeolite synthesis, the hydrothermal reactor used would also need to implement a form of agitation given the need to maintain a homogeneous mixture [55]. From the above analysis, Archimedes screw may be a favourable impeller during the hydrothermal synthesis due to its low rate of shearing; however, further research is required to ascertain this assumption. The synthesis gel that is used in the stage of hydrothermal synthesis varies considerably from the solution during aging. There is a need to understand the effect of agitation during hydrothermal synthesis. Homogenous mixing is critical if large-scale production is envisioned, a large-scale reactor would need to implement some form of agitation.

There have been significant studies on the use of class F fly ash for zeolite synthesis, which is pioneered Musyoka et al. [12, 56, 57]. The more recent studies have focused on using alternate techniques such as ultrasound to either used after fusing to decrease the hydrothermal synthesis time [12, 58], and the temperature needed in hydrothermal synthesis [58, 59] or complete replacement of the fusion process [60] A reduction in synthesis time was also noted by Musyoka et al. [61] when using ultrasonic synthesis using AMD and circumneutral mine waters in place of ultra-pure water. In addition, studies have also shown that ultrasound techniques can be used to investigate the formation mechanism of zeolites from coal fly ash using in-situ ultrasound techniques, namely of the zeolites A and X (zeolite X with novel hierarchal morphology) [57]. According to Murayama et al. [62], the three main steps involved in the mechanism of zeolite formation from coal fly ash are as follows:

- Dissolution of Si^{4+} and Al^{3+} in coal fly ash
- Condensation of silicate and aluminate ions in the alkali solution making the aluminosilicate gel
- Crystallisation of the aluminosilicate gel-forming zeolites

The above process can be represented pictorially in the following **Figure 7**.

Another technique used is microwave-assisted synthesis as reported by Querol et al. [64]. The activation time need during hydrothermal conversion was radically decreased from 24–48 hours to 30 minutes. It was further explained by Inada et al. [65] that continuous microwave irradiation retards the formation of zeolite in crystalline form, inhibiting the zeolite in the intermediate gel. It was found that early microwave irradiation enhances the zeolite formation. Heating during the middle stage, however, significantly inhibits zeolite formation. Therefore, it was found to be advisable to heat with microwave irradiation early on and then follow with conventional heating.

The feasibility though of scaling-up these alternate processes, such as microwave and ultrasonic-assisted synthesis, would prove challenging regarding the scale needed for large-scale zeolite synthesis. Ultrasound seems promising given its use of cavitation to produce high temperatures and pressures or "hot spots" under nearly ambient conditions [66]. The main form of cavitation being studied is acoustic cavitation generated by sound waves in which ultrasound generates, grows and collapses the cavities in microseconds. These hot spots can

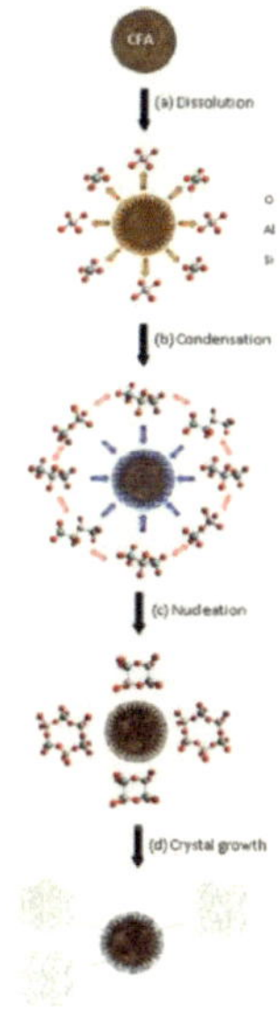

Figure 7. Illustration of the zeolite process of a coal fly ash particle [63].

produce temperatures greater than 5000 K and a cooling rate greater than 10^7 °C/s [60]. In a recent study by Ojumu et al. [60], the 90-min high-temperature fusion was completely replaced by a 10 minutes of high intensity ultrasound irradiation. It was shown that 24% of silicon was extracted from the fly ash, comparable to the 32% from fusion. It also reduced the crystallisation time of zeolite A and would be promising for up-scaling. On industrial scale, there is hardly any process carried out, which allows for up scale of the process. This is due to the lack of expertise in differing fields such as material science, acoustics, chemical engineering, etc. [66]. Hydrodynamic cavitation offers a cheaper alternative. It concerns generating cavitation from the flow of a liquid under controlled conditions through obstructions such as venture tubes and orifice plates. It is generated when the pressure at the throat falls below the vapour pressure of the liquid, the liquid then flashes and generates a number of cavities. When the pressure is recovered downstream of the mechanical constriction, these cavities then collapse [66–68]. The hydrodynamic cavitation is applied on a pilot plant scale within a jet loop reactor. In a study involving the scaling-up of a process to reduce sulphate concentrations within neutral mine drainage, a jet-loop reactor was used [69]. The study showed that the impingement and cavitation mixing techniques applied within the jet-loop reactor played an important role in enhancing the sulphate removal. This was due to the superior mixing in a jet-loop reactor compared to the laboratory-scale experiments that were also performed. The feasibility of scaling up a jet-loop pilot plant to full-scale seems more easily achieved than what would be needed to scale-up the high-intensity ultrasound process. Therefore, it may be necessary to investigate the application of a jet-loop reacting system incorporating fly ash and sodium hydroxide to ascertain the dissolution rates of the silicon and aluminium that can be achieved. A recent study by Nyale et al. [27] showed that the intense mixing within the jet-loop reactor increased the dissolution of the amorphous content of the fly ash used in the production of geopolymer from a slurry obtained from a jet-loop system. This may prove favourable for zeolite synthesis; however, systematic studies that investigate the potential to use jet-loop reactor as a process to completely replace the high-temperature fusion need to be conducted. This would involve comparison of jet-loop reacting system to the recent studies of involving ultrasound utilisation to determine which process might be more lucrative to up scale.

5. Up-scaling zeolite synthesis from coal fly ash

Querol et al. [48] ran experiments at pilot scale using a 10 m^3 R-410-A reactor vessel; the optimum conditions were generated from laboratory-scale experiments using Parr digestion bombs: they were found to be 2 M NaOH solution, solution/fly ash ratio of 21 kg^{-1}, $T = 150$°C, and time of 24 hours. It was concluded that the optimisation of synthesis yields would have to be specific for each type of fly ash from the differing power plants due to their difference in mineralogical and chemical compositions. There were reproducible results, however, in synthesising zeolite NaP1 from a specific power station fly ash. Also, high cation-exchange capacities (CEC) were reported for ammonium and heavy metals for the varying zeolites achieved.

Moriyama et al. [70] compared the CEC values of a conventional method of synthesis to a test unit and a pilot plant. The conventional method conditions were a liquid–solid ratio of 8 dm^3/kg, a temperature of 373 K and a holding time of 24 hours. The test unit consisted of a high-pressure kneader with a volume of 5 dm^3 fitted with a hot oil jacket around the pressure vessel. It was equipped with two sigma-shaped blades for kneading and a steam exhaust line to remove the water. The pilot plant was also a high-pressure kneader but at the volume of 0.6 m^3. This process had benefits of water removal during operation eliminating wastewater treatment. It was concluded that the influence of the pressure on the CEC values obtained depends on the original coal fly ash characteristics and the type of zeolite formed was gismodine (GIS) type.

Mainganye et al. [54] investigated the effect of the impeller design concerning agitation during the ageing step of zeolite NaP1 synthesis. Three impellers were tested (four flat blade, anchor and Archimedes screw) at varying agitation speeds (150, 200 and 300 rpm). The conditions were 48 hours of aging at 47°C and static hydrothermal treatment of 48 hours at 140°C. It was shown that the phase purity of the zeolite was strongly affected by the agitation rate and type of impeller used during the ageing step of synthesis. The optimal impeller was the four-blade impeller at an agitation rate of 200 rpm under the previous conditions mentioned, with a product yield of 0.98 ± 0.05 g zeolite/g fly ash and a space time yield of 15 ± 0.4 kg $d^{-1}m^{-3}$.

Du Plessis et al. [71] developed waste minimisation protocols for zeolite synthesis from coal fly ash. The study effectively constructed two protocols for the minimisation of the waste generated in zeolite synthesis from South African coal fly ash. It was shown that to successfully synthesise zeolites with the waste solution, there needed to be a pH adjustment giving the opportunity to recycle 40% if the supernatant waste back in to the system. With this pH adjustment, zeolites NaP1 and analcime were successfully synthesised. Due to a high Si/Al ratio in the waste after reusing the supernatant, the more predominant phase was zeolite analcime. It was also shown that it is possible to recycle 100% of the supernatant waste without adjusting the alkalinity but by adjusting the basic synthesis process slightly.

Du Plessis et al. [72] determine the distributional fate of elements during two of the most prominent zeolite synthesis methods from South African coal fly ash, namely the fusion process and the two-step process. An overall elemental material balance was performed around the two process routes. The results indicated that in the two-step method, almost all the elements were concentrated in the solid zeolite product, while in the fusion-assisted process, the elements mainly report to the solid waste. It was shown that toxic elements such as Pb, Hg, Al, As and Nb were found in both the supernatant waste and washing water within both processes.

Wdowin et al. [55] developed a technological line for converting coal fly ash to zeolites on a pilot plant scale. It could be divided into four stages: a reactor loading stage; a reaction stage; separation of the reaction products stage; and a final stage for processing the material obtained. The total volume of the reaction tank was 130 dm^3 (with a working volume of 100 dm^3). The tank was equipped with three 2-kW heaters, a temperature and tank filling probe and a stirrer. The stirrer was responsible for the homogenisation of the material and prevented aggregation of the material during the reaction process. The zeolite NaP1 was synthesised with a zeolite content of 81% under the conditions of: 20 kg fly ash; 12 kg NaOH; 90 dm^3 of H_2O; temperature of 80°C; and duration of 36 hours. The line was fully automated and was able to synthesise zeolites Na-X and Linde-A after changing of the synthesis conditions.

6. Future outlook towards up-scaling zeolite synthesis from coal fly ash

As can be seen from majority of the literature regarding zeolite up-scale synthesis from coal fly ash, the main zeolite being investigated is Na-P1. It is proposed that future investigation should focus on whether the faujasite zeolite (zeolite X) can be synthesised and targeted for large-scale production. Zeolite X is important, being the most widely employed zeolites on industrial scale and used extensively as a fluid catalytic cracking catalyst for refining oil and as materials for adsorbing and removing gaseous emissions [73]. In view of the recent study by Musyoka [12] on the synthesis of zeolite X with novel hierarchical morphology from the clear extract of fused fly ash with NaOH, a significant breakthrough would be to target the zeolite with such novel morphology for up scale. The hierarchical pore structure enables the zeolites to have maximum structural functions in a limited space and volume, thus conferring a high degree of diffusion efficiency [73]. This synthesis route would be much more advantageous over the complicated procedures reported in literature to obtain the hierarchical structured zeolites.

The hydrodynamic cavitation provided in jet-loop reacting system should be explored as suitable alternative to replace the high-temperature fusion process for a large-scale production process. Following from Nyale [27], who showed that the jet loop increased the amorphous content within the fly ash and reduced the amount of quartz and mullite, which would favour the synthesis of zeolites. It is reasoned that the same method could therefore be used for zeolite synthesis in the replacement of the fusion pre-synthesis. Although in Nyale's study, the goal was to synthesise geopolymers. However, zeolites and geopolymers are very similar regarding their atomic structures of silicon and aluminium atoms. Also, the jet-loop geopolymer produced leached significant amount of silicon which could be used for zeolite synthesis after a curing/leaching process of the slurry produced by the jet loop. This in turn could increase the yield of zeolite by creating a filtrate concentrated in the necessary silicon and aluminium.

Agitation is critical in the up scale of zeolite synthesis from coal fly ash; however, the challenge is to determine the effect of the impeller type and agitation during the hydrothermal treatment stage of the process. Although it has been shown that impeller type and agitation rate have a remarkable effect during the ageing step of zeolite synthesis [54], there is need to investigate these effects during the crystallisation process. It is postulated that impeller type that provide minimum shearing effect would be suitable during the hydrothermal process. At an industrial scale, stirring would need to be required not only to ensure homogeneity but also to ensure uniform heat distribution necessary for the crystallisation step. There have not been in depth studies to show the effect of agitation during the hydrothermal treatment process. Although the recent agitation used by Wdowin et al. [55] to synthesise zeolites from coal fly ash recorded 81% pure phase product, the purity could be due to the shear stress effect of the mechanical stirrer during the hydrothermal treatment process [54]. It is also not clear whether a pure zeolite would have been obtained if the clear filtrate was used instead of the total slurry. The aforementioned highlights the need for an in-depth studies into agitation during the hydrothermal process. The results of such studies might hold the promise for a large-scale synthesis that provides the highest and most pure yield of zeolite.

7. Conclusion

This review highlighted the present status and future outlook with respect to up-scaling of zeolite synthesis from coal fly ash. Although current studies have demonstrated that the seemingly energy-intensive fusion pre-synthesis step can be avoided by using the acoustic cavitation technique provided by ultrasound devices, the scalability of this device for large-scale production remains a subject of debate. The hydrodynamic cavitation provided by a jet-loop reacting system may be a suitable alternative. However, there is a need for systematic studies that would investigate the potential of jet-loop system, in comparison with the fusion and ultrasound pre-synthesis steps, for the dissolution of the silicon and aluminium from the coal fly ash. The critical need for agitation and suitable impeller type, during the hydrothermal synthesis on a large scale, was outlined with an emphasis on the requirement for minimum shearing during crystallisation in large hydrothermal stirred reactors. It is expected that these recommendations would provide understanding towards the design of an appropriate scale up operation for zeolite synthesis from coal fly ash.

Author details

James Philip Brassell[1], Tunde Victor Ojumu[1*] and Leslie Felicia Petrik[2]

*Address all correspondence to: ojumut@cput.ac.za

1 Department of Chemical Engineering, Cape Peninsula University of Technology, Cape Town, South Africa

2 Environmental Nano Sciences, Department of Chemistry, University of the Western Cape, Bellville, South Africa

References

[1] Holler, H. and U. Wirsching, Zeolite formation from fly-ash. Fortschritte der Mineralogie, 1985. 63(1): p. 21–43.

[2] Querol, X., N. Moreno, J.C. Umaa, A. Alastuey, E. Hernández, a. López-Soler, and F. Plana, Synthesis of zeolites from coal fly ash: an overview. International Journal of Coal Geology, 2002. 50: p. 413–423. DOI: 10.1016/S0166-5162(02)00124-6.

[3] Yao, Z.T., X.S. Ji, P.K. Sarker, J.H. Tang, L.Q. Ge, M.S. Xia, and Y.Q. Xi, A comprehensive review on the applications of coal fly ash. Earth-Science Reviews, 2015. 141: p. 105–121. DOI: 10.1016/j.earscirev.2014.11.016.

[4] Eskom. Integrated Report 2011 [Internet]. 2011. Available from: http://financialresults.co.za/2011/eskom_ar2011/ [Accessed: 2015-02-17]

[5] Nyale, S.M., C.P. Eze, R.O. Akinyeye, W.M. Gitari, S.A. Akinyemi, O.O. Fatoba, and L.F. Petrik, The leaching behaviour and geochemical fractionation of trace elements in hydraulically disposed weathered coal fly ash. Journal of environmental science and health. Part A, Toxic/hazardous Substances & Environmental Engineering, 2014. 49(October 2013): p. 233–242. DOI: 10.1080/10934529.2013.838929.

[6] Borm, P.J., Toxicity and occupational health hazards of coal fly ash (CFA). A review of data and comparison to coal mine dust. The Annals of Occupational Hygiene, 1997. 41(6): p. 659–676. DOI: 10.1016/S0003-4878(97)00026-4.

[7] Du Plessis, P.W., Process design for the up-scale zeolite synthesis from South African coal fly ash [thesis]. Cape Town: Cape Peninsula University of Technology; 2014.

[8] Mainganye, D., Synthesis of zeolites from South African coal fly ash: Investigation of scale-up conditions. [thesis]. Cape Town: Cape Peninsula University of Technology; 2012.

[9] Ahmaruzzaman, M., A review on the utilization of fly ash. Progress in Energy and Combustion Science, 2010. 36(3): p. 327–363. DOI: 10.1016/j.pecs.2009.11.003.

[10] Blissett, R.S. and N.A. Rowson, A review of the multi-component utilisation of coal fly ash. Fuel, 2012. 97: p. 1–23. DOI: 10.1016/j.fuel.2012.03.024.

[11] Vassilev, S.V. and C.G. Vassileva, A new approach for the classification of coal fly ashes based on their origin, composition, properties, and behaviour. Fuel, 2007. 86(10–11): p. 1490–1512. DOI: 10.1016/j.fuel.2006.11.020.

[12] Musyoka, N.M., Zeolite A, X and Cancrinite from South African coal fly ash: mechanism of crystallization, routes to rapid synthesis and new morphology [thesis]. Cape Town: University of the Western Cape; 2012.

[13] Kutchko, B. and A. Kim, Fly ash characterization by SEM–EDS. Fuel, 2006. 85(17–18): p. 2537–2544. DOI: 10.1016/j.fuel.2006.05.016.

[14] Fisher, G.L., B.A. Prentice, and D. Silberman, Physical and morphological studies of size-classified coal fly ash. Environmental Science and Technology, 1978. 12(4): p. 447–451. DOI: 10.1021/es60140a008.

[15] Wadge, A. and M. Hutton, The leachability and chemical speciation of selected trace elements in fly ash from coal combustion and refuse incineration. Environmental Pollution, 1987. 48(2): p. 85–99. DOI: 10.1016/0269-7491(87)90089-3.

[16] Khanra, S., D. Mallick, S.N. Dutta, and S.K. Chaudhuri, Studies on the phase mineralogy and leaching characteristics mineralogy and leaching characteristics Pollution, 1998. 107(1-4): p. 251–275. DOI: 10.1023/A:1004947519170.

[17] Wang, Y., D. Ren, and F. Zhao, Comparative leaching experiments for trace elements in raw coal, laboratory ash, fly ash and bottom ash. International Journal of Coal Geology, 1999. 40(2–3): p. 103–108. DOI: 10.1016/S0166-5162(98)00062-7.

[18] Choi, S.K., S. Lee, Y.K. Song, and H.S. Moon, Leaching characteristics of selected Korean fly ashes and its implications for the groundwater composition near the ash disposal mound. Fuel, 2002. 81(8): p. 1083–1090. DOI: 10.1016/S0016-2361(02)00006-6.

[19] Praharaj, T., M.A. Powell, B.R. Hart, and S. Tripathy, Leachability of elements from subbituminous coal fly ash from India. Environment International, 2002. 27(8): p. 609–615. DOI: 10.1016/S0160-4120(01)00118-0.

[20] Ugurlu, A., Leaching characteristics of fly ash. Environmental Geology, 2004. 46: p. 890–895. DOI: 10.1007/s00254-004-1100-6.

[21] Jegadeesan, G., S.R. Al-Abed, and P. Pinto, Influence of trace metal distribution on its leachability from coal fly ash. Fuel, 2008. 87(10–11): p. 1887–1893. DOI: 10.1016/j.fuel.2007.12.007.

[22] Dutta, B.K., S. Khanra, and D. Mallick, Leaching of elements from coal fly ash: assessment of its potential for use in filling abandoned coal mines. Fuel, 2009. 88(7): p. 1314–1323. DOI: 10.1016/j.fuel.2009.01.005.

[23] Izquierdo, M. and X. Querol, Leaching behaviour of elements from coal combustion fly ash: an overview. International Journal of Coal Geology, 2012. 94: p. 54–66. DOI: 10.1016/j.coal.2011.10.006.

[24] Xiang, W., B. Han, D. Zhou, and A. Nzihou, Physicochemical properties and heavy metals leachability of fly ash from coal-fired power plant. International Journal of Mining Science and Technology, 2012. 22(3): p. 405–409. DOI: 10.1016/j.ijmst.2011.12.002.

[25] Belviso, C., F. Cavalcante, S. Di Gennaro, A. Palma, P. Ragone, and S. Fiore, Mobility of trace elements in fly ash and in zeolitised coal fly ash. Fuel, 2015. 144(0): p. 369–379. DOI: http://dx.doi.org/10.1016/j.fuel.2014.12.037.

[26] Komonweeraket, K., B. Cetin, A.H. Aydilek, C.H. Benson, and T.B. Edil, Effects of pH on the leaching mechanisms of elements from fly ash mixed soils. Fuel, 2015. 140: p. 788–802. DOI: 10.1016/j.fuel.2014.09.068.

[27] Nyale, S.M., Geopolymers from South African fly ash: synthesis and characteristics. [thesis]. Cape Town: University of the Western Cape; 2014.

[28] Cundy, C.S. and P.A. Cox, The hydrothermal synthesis of zeolites: history and development from the earliest days to the present time. Chemical Reviews, 2003. 103: p. 663–701. DOI: 10.1021/cr020060i.

[29] Xu, R., W. Pang, J. Yu, Q. Huo, and J. Chen, Chemistry of zeolites and related porous materials: synthesis and structure. Singapore: John Wiley & Sons; 2007, 696p. DOI: 10.1002/9780470822371.

[30] Jha, B. and D.N. Singh, A review on synthesis, characterization and industrial applications of flyash zeolites. Journal of Materials Education, 2011. 33: p. 65–132. DOI: 10.1002/chin.201225227.

[31] Auerbach, S.M., K.A. Carrado, and P.K. Dutta, Handbook of zeolite science and technology. New York: Marcel Dekker; 2003, 1204p. DOI: 10.1201/9780203911167.

[32] Shigemoto, N., H. Hayashi, and K. Miyaura, Selective formation of Na-X zeolite from coal fly ash by fusion with sodium hydroxide prior to hydrothermal reaction. Journal of Materials Science, 1993. 28: p. 4781–4786. DOI: 10.1007/BF00414272.

[33] Zhao, X.S., G.Q. Lu, and H.Y. Zhu, Effects of ageing and seeding on the formation of zeolite Y from coal fly ash. Journal of Porous Materials, 1997. 4(4): p. 245–251. DOI: 10.1023/A:1009669104923.

[34] Steenbruggen, G. and G.G. Hollman, The synthesis of zeolites from fly ash and the properties of the zeolite products. Journal of Geochemical Exploration, 1998. 62: p. 305–309. DOI: 10.1016/S0375-6742(97)00066-6.

[35] Hollman, G.G., G. Steenbruggen, and M. Janssen-Jurkovičová, A two-step process for the synthesis of zeolites from coal fly ash. Fuel, 1999. 78: p. 1225–1230. DOI: 10.1016/S0016-2361(99)00030-7.

[36] Murayama, N., H. Yamamoto, and J. Shibata, Zeolite synthesis from coal fly ash by hydrothermal reaction using various alkali sources. Journal of Chemical Technology and Biotechnology, 2002. 77(June 2001): p. 280–286. DOI: 10.1002/jctb.604.

[37] Ojha, K., N.C. Pradhan, and A.N. Samanta, Zeolite from fly ash: synthesis and characterization. Bulletin of Materials Science, 2004. 27(6): p. 555–564. DOI: 10.1007/BF02707285.

[38] Inada, M., Y. Eguchi, N. Enomoto, and J. Hojo, Synthesis of zeolite from coal fly ashes with different silica-alumina composition. Fuel, 2005. 84: p. 299–304. DOI: 10.1016/j.fuel.2004.08.012.

[39] Rungsuk, D., R. Apiratikul, V. Pavarajarn, and P. Pavasant, Zeolite synthesis from fly ash from coal-fired power plant by fusion method. In: Proceedings of the 2nd joint international conference on "Sustainable Energy and Environment (SEE 2006)" 21–23 November 2006; Bangkok, Thailand. 2006. p. 2–6.

[40] Rayalu, S., S.U. Meshram, and M.Z. Hasan, Highly crystalline faujasitic zeolites from flyash. Journal of Hazardous Materials, 2000. 77(1): p. 123–131. DOI: 10.1016/S03043894(00)00212-0.

[41] Tanaka, H., S. Matsumura, S. Furusawa, and R. Hino, Conversion of coal fly ash to Na-X zeolites. Journal of Materials Science Letters, 2003. 22(c): p. 323–325. DOI: 10.1023/A:1022329002370.

[42] Belviso, C., F. Cavalcante, and S. Fiore, Synthesis of zeolite from Italian coal fly ash: differences in crystallization temperature using seawater instead of distilled water. Waste Management, 2010. 30(5): p. 839–847. DOI: 10.1016/j.wasman.2009.11.015.

[43] IZA. IZA Synthesis Commission [Internet]. 2013. Available from: http://www.iza-online.org/synthesis/default.htm [Accessed: 2015-08-03]

[44] Molina, A. and C. Poole, A comparative study using two methods to produce zeolites from fly ash. Minerals Engineering, 2004. 17(2): p. 167–173. DOI: http://dx.doi.org/10.1016/j.mineng.2003.10.025.

[45] Querol, X., F. Plana, A. Alastuey, and A. López-Soler, Synthesis of Na-zeolites from fly ash. Fuel, 1997. 76(8): p. 793–799. DOI: 10.1016/S0016-2361(96)00188-3.

[46] Querol, X., A. Alastuey, J. Fernández-Turiel, and A. López-Soler, Synthesis of zeolites by alkaline activation of ferro-aluminous fly ash. Fuel, 1995. 74(8): p. 1226–1231. DOI: 10.1016/0016-2361(95)00044-6.

[47] Ríos, R., A. Carlos, C.D. Williams, and C.L. Roberts, A comparative study of two methods for the synthesis of fly ash-based sodium and potassium type zeolites. Fuel, 2009. 88(8): p. 1403–1416. DOI: 10.1016/j.fuel.2009.02.012.

[48] Querol, X., J.C. Umana, F. Plana, A. Alastuey, A. Lopez-soler, A. Medinaceli, A. Valero, M.J. Domingo, and E. Garcia-rojo, Synthesis of zeolites from fly ash at pilot plant scale. Examples of potential applications. Fuel, 2001. 80: p. 857–865. DOI: 10.1016/S0016-2361(00)00156-3.

[49] Musyoka, N.M., L.F. Petrik, O.O. Fatoba, and E. Hums, Synthesis of zeolites from coal fly ash using mine waters. Minerals Engineering, 2013. 53: p. 9–15. DOI: 10.1016/j.mineng.2013.06.019.

[50] Belviso, C., F. Cavalcante, F. Javier Huertas, A. Lettino, P. Ragone, and S. Fiore, The crystallisation of zeolite (X- and A-type) from fly ash at 25°C in artificial sea water. Microporous and Mesoporous Materials, 2012. 162: p. 115–121. DOI: 10.1016/j.micromeso.2012.06.028.

[51] Boycheva, S., D. Zgureva, and A. Shoumkova, Recycling of lignite coal fly ash by its conversion into zeolites. Coal Combustion and Gasification Products, 2014. 6: p. 1–8. DOI: 10.4177/CCGP-D-14-00008.1.

[52] Casci, J.L., Zeolite molecular sieves: preparation and scale-up. Microporous and Mesoporous Materials, 2005. 82(3): p. 217–226. DOI: 10.1016/j.micromeso.2005.01.035.

[53] Marrot, B., C. Bebon, D. Colson, and J.P. Klein, Influence of the shear rate during the synthesis of zeolites. Crystal Research and Technology, 2001. 36(figure 1): p. 269–281. DOI: 10.1002/1521-4079(200103)36:3<269::AID-CRAT269>3.0.CO;2-Q.

[54] Mainganye, D., T.V. Ojumu, and L. Petrik, Synthesis of zeolites Na-P1 from South African coal fly ash: effect of impeller design and agitation. Materials, 2013. 6: p. 2704–2089. DOI: 10.3390/ma6052074.

[55] Wdowin, M., M. Franus, R. Panek, L. Badura, and W. Franus, The conversion technology of fly ash into zeolites. Clean Technologies and Environmental Policy, 2014. 16: p. 1217–1223. DOI: 10.1007/s10098-014-0719-6.

[56] Musyoka, N.M., Hydrothermal synthesis and optimisation of zeolite Na-P1 from South African coal fly ash [thesis]. Cape Town: University of the Western Cape; 2009.

[57] Musyoka, N.M., L.F. Petrik, E. Hums, H. Baser, and W. Schwieger, In situ ultrasonic diagnostic of zeolite X crystallization with novel (hierarchical) morphology from coal fly ash. Ultrasonics, 2014. 54(2): p. 537–543. DOI: 10.1016/j.ultras.2013.08.005.

[58] Bukhari, S.S., S. Rohani, and H. Kazemian, Effect of ultrasound energy on the zeolitization of chemical extracts from fused coal fly ash. Ultrasonics sonochemistry, 2016. 28: p. 47–53. DOI: 10.1016/j.ultsonch.2015.06.031.

[59] Belviso, C., F. Cavalcante, A. Lettino, and S. Fiore, Effects of ultrasonic treatment on zeolite synthesized from coal fly ash. Ultrasonics Sonochemistry, 2011. 18(2): p. 661–668. DOI: 10.1016/j.ultsonch.2010.08.011.

[60] Ojumu, T., P. Du Plessis, and L. Petrik, Synthesis of zeolite A from coal fly ash using ultrasonic treatment-a replacement for fusion step. Ultrasonics Sonochemistry, 2016. 31: p. 342–349. DOI: 10.1016/j.ultsonch.2016.01.016.

[61] Musyoka, N.M., L.F. Petrik, and E. Hums, Ultrasonic assisted synthesis of zeolite A from coal fly ash using mine waters (acid mine drainage and circumneutral mine water) as a substitute for ultra pure water. In: Rüde, R.T., Freund, A. & Wolkersdorfer, Ch. (editors): Mine water – managing the challenges; Aachen, Germany. 2011. p. 423–428.

[62] Murayama, N., H. Yamamoto, and J. Shibata, Mechanism of zeolite synthesis from coal fly ash by alkali hydrothermal reaction. International Journal of Mineral Processing, 2002. 64: p. 1–17. DOI: 10.1016/S0301-7516(01)00046-1.

[63] Bukhari, S.S., J. Behin, H. Kazemian, and S. Rohani, Conversion of coal fly ash to zeolite utilizing microwave and ultrasound energies: a review. Fuel, 2015. 140: p. 250–266. DOI: 10.1016/j.fuel.2014.09.077.

[64] Querol, X., A. Alastuey, A. López-Soler, F. Plana, J.M. Andrés, R. Juan, P. Ferrer, and C.R. Ruiz, A fast method for recycling fly ash: microwave-assisted zeolite synthesis. Environmental Science and Technology, 1997. 31(9): p. 2527–2533. DOI: 10.1021/es960937t.

[65] Inada, M., H. Tsujimoto, Y. Eguchi, N. Enomoto, and J. Hojo, Microwave-assisted zeolite synthesis from coal fly ash in hydrothermal process. Fuel, 2005. 84: p. 1482–1486. DOI: 10.1016/j.fuel.2005.02.002.

[66] Moholkar, V.S., P. Senthil Kumar, and A.B. Pandit, Hydrodynamic cavitation for sonochemical effects. Ultrasonics Sonochemistry, 1999. 6(1–2): p. 53–65. DOI: 10.1016/S1350-4177(98)00030-3.

[67] Jyoti, K.K. and A.B. Pandit, Water disinfection by acoustic and hydrodynamic cavitation. Biochemical Engineering Journal, 2001. 7(3): p. 201–212. DOI: 10.1016/S1369-703X(00)00128-5.

[68] Gogate, P.R. and A.B. Pandit, A review and assessment of hydrodynamic cavitation as a technology for the future. Ultrasonics sonochemistry, 2005. 12(1–2): p. 21–27. DOI: 10.1016/j.ultsonch.2004.03.007.

[69] Madzivire, G., W.M. Gitari, V.R.K. Vadapalli, and L.F. Petrik, Jet loop reactor application for mine water treatment using fly ash, lime and aluminium hydroxide. Interna-

tional Journal of Environmental Science and Technology, 2013. 12: p. 173–182. DOI: 10.1007/s13762-013-0417-7.

[70] Moriyama, R., S. Takeda, M. Onozaki, Y. Katayama, K. Shiota, T. Fukuda, H. Sugihara, and Y. Tani, Large-scale synthesis of artificial zeolite from coal fly ash with a small charge of alkaline solution. Fuel, 2005. 84: p. 1455–1461. DOI: 10.1016/j.fuel.2005.02.026.

[71] Du Plessis, P.W., T.V. Ojumu, and L.F. Petrik, Waste minimization protocols for the process of synthesizing zeolites from South African coal fly ash. Materials, 2013. 6: p. 1688–1703. DOI: 10.3390/ma6051688.

[72] Du Plessis, P., T. Ojumu, O. Fatoba, R. Akinyeye, and L. Petrik, Distributional fate of elements during the synthesis of zeolites from South African coal fly ash. Materials, 2014. 7: p. 3305–3318. DOI: 10.3390/ma7043305.

[73] Liu, B., F. Chen, L. Zheng, J. Ge, H. Xi, and Y. Qian, Synthesis and structural properties of hierarchically structured aluminosilicates with zeolite Y (FAU). RSC Advances, 2013. 3: p. 15075–15084. DOI: 10.1039/c3ra41862f.

Chapter 2

Hydrothermal Conversion of South African Coal Fly Ash into Pure Phase Zeolite Na-P1

Mugera W. Gitari, Leslie F. Petrik and
Nicholas M. Musyoka

Additional information is available at the end of the chapter

http://dx.doi.org/10.5772/63399

Abstract

South African coal combustion power utilities generate huge amounts of coal fly ash that can be beneficiated into zeolitic products. This chapter reports on the optimization of the presynthesis and synthesis conditions for a pure-phase zeolite Na-P1 from selected South African coal fly ashes. The hydrothermal treatment time, temperature, and molar quantities of water during the hydrothermal treatment step were successfully optimized. The optimum hydrothermal treatment time and temperature were 48 h and 140°C, respectively. Pure-phase zeolite Na-P1 was obtained with a molar regime of 1 SiO_2:0.36 Al_2O_3:0.59 NaOH:0.49 H_2O at an aging temperature of 47°C for 48 h. The optimized conditions were applied to two fly ashes from two coal-fired power utilities, and high-purity zeolite Na-P1 was obtained. The third coal fly ash with a different chemical composition gave a low-quality Na-P1 under the optimized conditions. The cation exchange capacity for the high-purity zeolite phase was 4.11 mEq/g, indicating that the adjustment of reactant composition and presynthesis or synthesis parameters leads to yields of high-quality zeolite Na-P1. The results also show that conversion of the coal fly ash into high-purity zeolite also depends on the chemical and mineralogical composition of the coal fly ash.

Keywords: zeolite Na-P1, coal fly ash, hydrothermal synthesis, optimization, aging step, CEC

1. Introduction

South Africa relies mainly on coal combustion for power generation [1]. Low-grade bituminous coal is combusted for power generation; this in turn generates huge volumes of waste materials such as bottom ash, fly ash, boiler slag, flue gas desulfurization sludge, and noncaptured particles [1, 2]. Of most concern is fly ash, which is mainly collected from flue gases by means of mechanical devices [3, 4]. After collection, fly ash is hydraulically transported to holding ponds, lagoons, landfills, and slag heaps, where it can be reacquired for treatment purposes or discarded or conveyed to ash heaps in dry disposal systems. The disposal and management of the huge quantities of coal fly ash has been a concern to industrial environmental managers and the scientific community and are in constant search for bulky beneficial utilization of coal fly ash to offset management costs and protect the environment. Approximately 27 Mt of fly ash is produced annually in South African power utilities [1]. Of this, only 5% is utilized beneficially. The remainder is discarded in surface impoundments, such as ash retention dams, where it continues to evolve chemically and mineralogically with possible environmental impacts [5]. Several researchers have proposed strategies for bulky utilization of fly ash, which include (i) applications as an additive in the manufacturing of cement, concrete, construction materials, and road pavements [6, 7], (ii) utilization of fly ash in zeolite synthesis for wastewater treatment [8–14], and (iii) use of fly ash in the neutralization of acid mine drainage and mine backfill [1, 15]. Most of these approaches are an attempt to beneficially use coal fly ash to supplement the cost of disposal and management and to reduce the negative environmental impact.

The application of coal fly ash in zeolite synthesis is of major interest due to the many industrial applications of zeolites, such as catalysis or catalyst carriers, adsorbents for the removal of inorganic and organic contaminants from wastewaters, management of radioactive wastes, gas separation, slow-release fertilizer, and in the manufacture of detergents [14]. The conversion of these low-cost waste products into products of higher value such as zeolites would allow the beneficiation of fly ash in an environment-friendly condition in addition to economic benefits. South African class F coal fly ash has been confirmed to be a good feedstock of Al and Si for zeolite synthesis because of its compositional dominance of aluminosilicate and silicate phases [14, 16]. Zeolite synthesis from fly ash is one of the potential environmentally useful applications of fly ash to produce high-value industrial products [10]. The major potential applications of zeolites synthesized from fly ash are based on their use as high capacity ion exchangers in industrial water waste treatment due to their large pore volumes [8, 10, 17]. Zeolites have also proven to be good candidates for use in soil decontamination [18] and have also shown great potential for use in the removal of postcombustion gases such as SO_x and NO_x [19].

Through various synthesis methods, researchers have synthesized various types of zeolites from fly ash, such as zeolite Na-P1 [20], zeolite A [21, 22], and zeolite ZSM-5 [23]. However, few studies have been successful in the conversion of fly ash into pure-phase zeolites [24]. Querol et al. [13] evaluated the synthesis of zeolites from fly ash at pilot scale using various Spanish fly ashes. The authors observed that, to obtain pure-phase zeolites, the optimum synthesis conditions have to be established for each coal fly ash. They attributed this to differences in mineralogical and chemical composition. South African coal fly ashes differ significantly in the chemical and mineralogical composition and this formed the basis of this work, as they have not been evaluated before as feedstocks to synthesize pure-phase zeolites.

The zeolite Na-P1 was chosen as the model zeolite to test our hypothesis. Zeolite Na-P1 belongs to an important group of zeolites that can be synthesized under mild hydrothermal synthesis conditions without using templates, making them potentially economically viable and a green product. The narrow particle size distribution of the zeolite Na-P1 coupled with the micron-sized crystallites and the unusual framework flexibility gives this zeolite unique, favorable ion exchange and water sorption properties [25]. Very few studies have reported on the synthesis of pure-phase zeolites, although numerous studies have been carried out in the synthesis of zeolites from coal fly ash using modified synthesis procedures [13, 21, 26]. Most of the studies have yielded low-quality zeolites due to the incomplete conversion of the coal fly ash matrix into zeolitic phase. Successful transformation of South African fly ash into impure zeolite P using the mild temperature method (100°C) was achieved by Petrik et al. [27].

Szostak [28] classified the factors affecting the zeolite crystallization as composition of the reaction matrix, time of reaction, temperature, and history-dependent factors such as stirring, aging, nature of mixture, and order of mixing. Casci [29] observed that the variation of individual reaction components and reaction variables, such as reaction temperature, alkaline concentration, and time of synthesis, influence the type of zeolite synthesized, the quality of the zeolite, and also the efficiency of the synthesis process. To achieve the complete dissolution of coal fly ash matrix and conversion into the zeolitic phase, the modification of presynthesis steps and synthesis and postsynthesis conditions is recommended. A two-step synthesis procedure developed by Hollman et al. was adapted for our optimization procedures using South African coal fly ashes [30]. The optimized parameters were hydrothermal treatment time, temperature, and water content during hydrothermal treatment. This chapter reports on the successful conversion of South African coal fly ash into pure-phase Na-P1 zeolite.

2. Sampling and experimental procedures

2.1. Sample handling

Pulverized coal fly ashes were collected from the ash collection systems at three different coal power utilities in Mpumalanga, South Africa. These coal fly ashes were used as feedstock for the zeolite synthesis. The sampled coal fly ash was stored in tightly locked plastic containers to prevent the ingress of moisture and CO_2 and out of direct sunlight. Coal fly ash consists of metastable phases and reactive components, which were formed during high-temperature combustion, and they are likely to react with CO_2 and moisture in the atmosphere forming new products. This might alter the chemical and mineralogical composition of the coal fly ash [30].

2.2. Synthesis procedures and equipment

A two-step process for the synthesis of zeolites from fly ash adopted from Hollman et al. [11] was followed, whereby a mixture of coal fly ash and alkaline solution was subjected to (1) aging and (2) hydrothermal treatment.

The setup for the aging process is shown in **Figure 1**. In this aging step, coal fly ash was mixed with sodium hydroxide (NaOH) pellets in a ratio of 1:1. The NaOH pellets were first dissolved in 100 mL ultra-pure water in a separate beaker and then added to the coal fly ash in a 250 mL HDPC temperature-resistant sealable bottle, and a magnetic bar was added and the mixture in the sealed bottle was then heated on a magnetic stirrer hotplate. The hotplate was adjusted to a predetermined temperature of 47°C, which was controlled by a temperature probe, while the speed of rotation was set at 800 rpm. The aging temperature and time were kept constant at 47°C and 48 h, respectively, as reported from a previous study [27], to be optimum for the dissolution of aluminosilicate matrix of the coal fly ash in alkaline media.

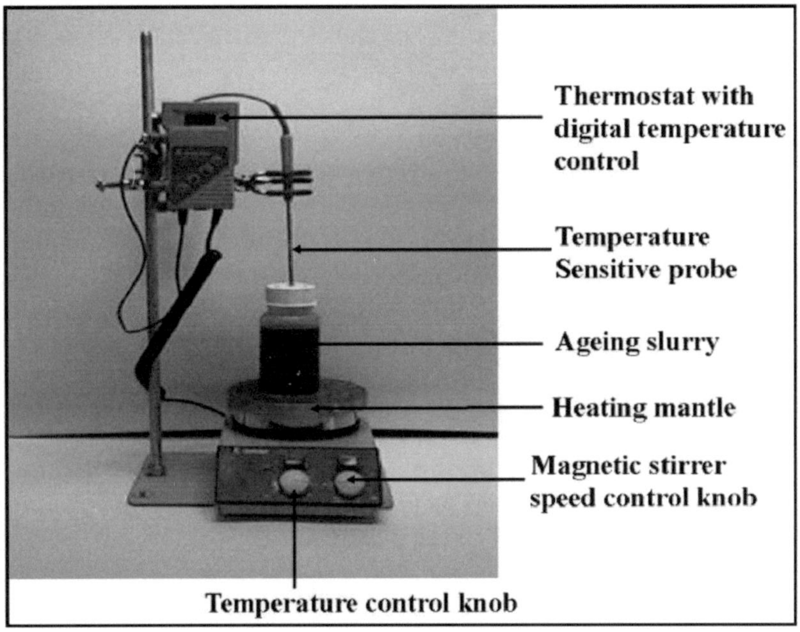

Figure 1. Experimental set-up for the aging process.

For hydrothermal treatment, varying amounts of ultra-pure water were added to the slurry after aging, the mixture was stirred, and the resulting homogenous solution was transferred in aliquots of 10 mL into a 23 mL Parr bomb. The crystallization of the feedstock was achieved by placing the mixture in sealed Parr bombs in a thermostated Memmert hot air oven for a predetermined time and temperature (100–160°C at intervals of 20°C with parallel time variation of between 12 and 48 h at intervals of 12 h).

After hydrothermal treatment at the predetermined time and temperature, the sample bearing Parr bombs were removed from the oven and allowed to cool down to room temperature. The cooled mixture was filtered to obtain the filtrates and the solid residue products. The pH of the filtrate was determined, whereas the solid product was washed thoroughly with ultra-pure water until the washing attained a pH of 9 to 10. The supernatant solution was acidified and kept refrigerated until analysis for chemical species. When the pH of the rinse solution had reached 9, the phases were separated and the solid product was recovered and dried overnight at 90°C and then transferred into airtight plastic containers before characterization. The pH and EC of solutions and supernatants were measured using HANNA HI 991301 portable pH/EC/TDS/temperature meter, which was calibrated before taking the measurements.

2.3. Characterization

The chemical composition of the coal fly ash and the synthesized zeolitic materials was carried out using X-ray fluorescence spectroscopy (XRF; Philips PW 1480 X-ray spectrometer). The samples were prepared by mixing 9 g of coal fly ash or zeolitic material with 2 g of a binder (10% C-wax binder and 90% EMU powder). The mixture was then thoroughly shaken, poured into the mould, and pelletized at a pressure of 15 tons for ~1 min using a Dickie and Stockler manual pelletizer. Loss on ignition (LOI) was measured by placing the samples in the furnace at 1000°C for at least 45 min. The instrument operating conditions for major element analysis were on a fused glass bead at 40 kV and 50 mA tube, whereas those for trace species were on a powder briquette at 50 kV and 40 mA tube operating conditions.

The samples for X-ray diffraction (XRD) were ground to a fine powder. The qualitative and quantitative XRD analyses of the coal fly ash and synthesized zeolite samples were done by placing the powder samples in a sample holder. The sample pattern was generated by a Philips X-ray diffractometer with Cu Kα radiation. The crystalline phases were identified by matching the obtained XRD profile with the powder diffraction file data (Joint Committee of Powder Diffraction Standards) files for inorganic compounds. The quantification of crystalline and amorphous phases was done as follows: After the addition of 20% Si (Aldrich; 99% purity) for the determination of amorphous content, the samples were milled in a McCrone micronizing mill. XRD profiles were acquired with a PANalytical X'Pert Pro powder diffractometer with X'Celerator detector and variable divergence and receiving slits with Fe filtered Co Kα radiation generated at 20 mA and 40 kV. The phases were identified using X'Pert Highscore Plus software. The relative phase amounts (wt.%) were estimated using the Rietveld method (Autoquan Program).

The morphology of the coal fly ash and the zeolitic products were examined using a scanning electron microscope (Hitachi X-650, Tokyo, Japan) with a compact detecting unit lead detector at 25 kV. The samples were mounted on aluminum pegs and sput coated with a thin film of gold for conductivity. For analysis by transmission electron microscopy (TEM), the samples were prepared by diluting a suspension of the synthesis products in ethanol, ultrasonicating, and depositing a drop onto S147-4 Holey carbon film 400 mesh Cu grids. A 200 kV field emission gunlens 1 was used with spotsize 3 at 200 kV using high-resolution TEM (HRTEM) Tecnai G2 F20 XTwin MAT.

The cation exchange capacity (CEC) was determined following the methods of Amrhein et al. [31]. The untreated coal fly ash was first saturated through three repeated rinsing steps with 1.0 M sodium acetate (pH 8.2) followed by four washings with ultra-pure water. The extraction of the exchangeable cations was done with three aliquots of 1.0 M ammonium acetate (pH 8.2). A 0.5 g of the zeolitic products was extracted in succession with four 25 mL aliquots of 1.0 M ammonium acetate (pH 8.2). On the addition of the 25 mL extracting solution, the mixture was then continuously agitated for 15 min, centrifuged for 15 min, and decanted. The procedure was repeated four times and the cumulative extract was collected for each sample. The concentrations of exchangeable cations (Na^+, Mg^{2+}, Ca^{2+}, and K^+) in the final solution were determined by inductively coupled plasma-atomic emission spectrometry. The CEC was then reported as mEq/g sample.

The surface area and pore size determination of the zeolitic products were carried out using the Brunauer-Emmett-Teller (N2-BET) surface analysis technique. The samples (0.35–0.5 g) were first outgassed at 110°C using helium gas. Micromeritics Tristar instrument (Tristar3000, Micromeritics, Norcross, GA, USA) was used with nitrogen as the analysis gas based on a 5 point with 30 adsorption and 30 desorption points.

3. Results and discussion

3.1. Characterization of coal fly ash and zeolitic products

The chemical composition of the coal fly ash feedstock and the zeolitic products reported as oxides is presented in **Table 1**.

Chemical component	Coal fly ash A (%, w/w)	Coal fly ash B (%, w/w)	Coal fly ash C (%, w/w)	Zeolite A (%, w/w)	Zeolite B (%, w/w)	Zeolite C (%, w/w)
SiO_2	50.91	49.79	54.92	36.69	35.16	33.95
Al_2O_3	30.91	31.75	27.27	25.17	22.57	31.81
Fe_2O_3	3.46	3.17	4.78	2.28	1.78	2.28
MnO	0.02	0.00	0.05	0.06	0.04	0.04
MgO	1.48	0.98	1.07	2.21	1.48	1.12
CaO	6.2	4.62	3.69	6.07	4.36	3.65
Na_2O	0.10	0.09	0.07	6.53	9.17	5.49
K_2O	0.60	0.63	0.66	0.13	0.13	0.14
P_2O_5	0.56	0.67	0.61	0.04	0.06	0.04
TiO_2	1.65	1.46	0.3	1.58	1.24	1.95
SO_3	0.24	0.23	1.71	0.06	0.06	0.03
LOI	3.85	6.59	4.44	11.92	13.91	10.99
Total	99.99	99.98	99.58	99.81	98.2	99.21
SiO_2/Al_20_3	1.65	1.57	2.01	1.45	1.56	1.07

Table 1. Chemical composition of coal fly ash and synthesized zeolites.

For all coal fly ash feedstock, the SiO_2+Al_2O_3+Fe_2O_3 is ≥70%, meaning the fly ashes can be classified as class F fly ash (ATSM method C 618). This is consistent with coal fly ash from the combustion of bituminous coal from South Africa [32]. The LOI ranged from 3.85% to 6.59% (w/w). LOI represents the unburned carbon in coal fly ash. The mean SiO_2/Al_2O_3 ratio ranged from 1.65 for coal fly ash A, 1.57 for B, and 2.01 for C. The SiO_2/Al_2O_3 ratio is important in that it directly governs both the Si/Al ratio of the zeolite product and the incorporation of Al in the framework structure. The presence of CaO and MgO plays a significant role in zeolite synthesis. The Ca^{2+} and Mg^{2-} ions act as competing cations during synthesis [33]. The lower content of CaO in coal fly ash C could result in the lower alkalinity of the solution during aging and hydrothermal synthesis, which could affect the rate of depolymerization and monomerization process [33]. The SiO_2/Al_2O_3 for zeolite A (from coal fly ash A) was 1.45 less than in the feedstock fly ash, indicating the inefficient conversion of the feedstock. Na_2O was observed to be higher in all the zeolites compared to the feedstock; this is because of the incorporation of Na as a charge-balancing cation, as NaOH was used as the alkaline coal fly ash matrix dissolution agent.

The qualitative XRD results of coal fly ash A are presented in **Figure 2** (note that the XRD profile is similar for coal fly ashes B and C and hence not shown), whereas **Figure 3** shows a summary of the quantitative analysis of the three coal fly ashes.

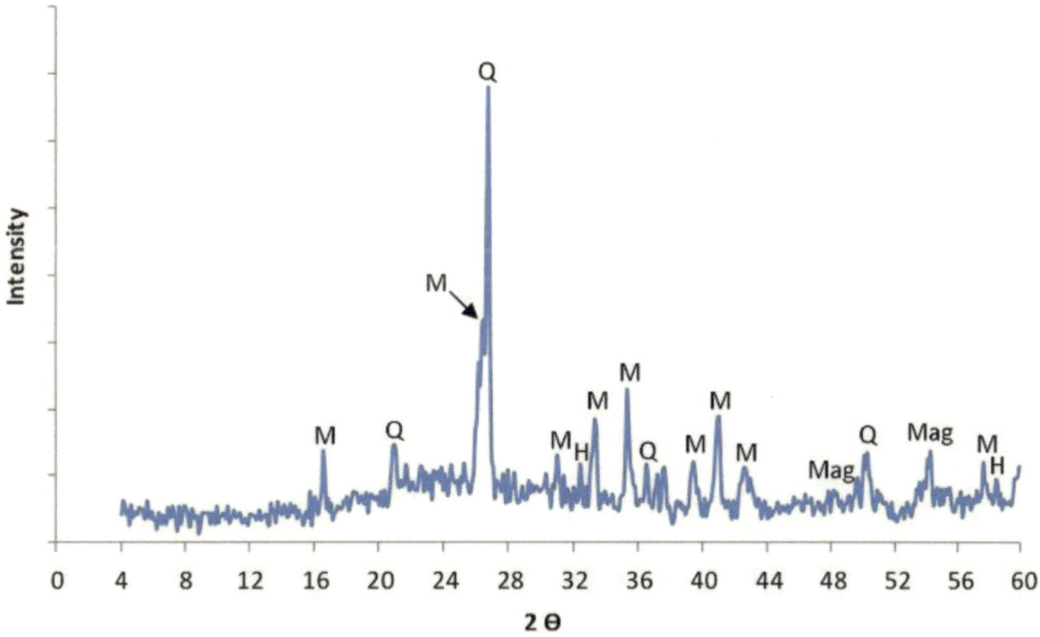

Figure 2. XRD profile of coal fly ash A (Q = quartz, M = mullite, H = hematite, Mag = magnetite).

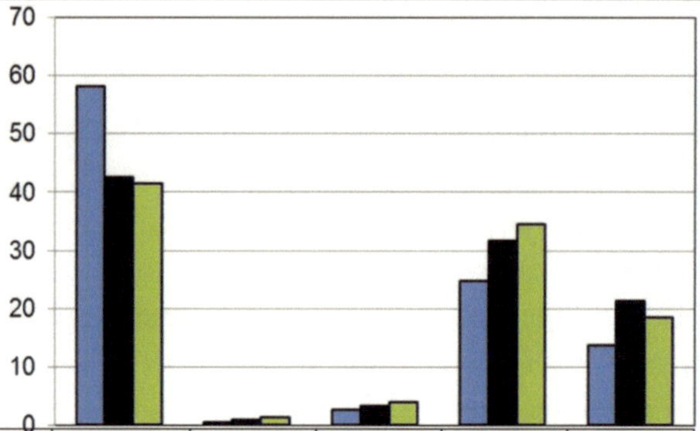

	Amorphous (%)	Hematite (%)	Magnetite (%)	Mullite (%)	Quartz (%)
Coal fly ash A	58.22	0.53	2.75	24.84	13.66
Coal fly ash B	42.60	1.06	3.24	31.72	21.38
Coal fly ash C	41.55	1.39	3.90	34.55	18.61

Figure 3. Relative quantitative XRD analysis of the three coal fly ashes.

The qualitative XRD analysis indicated the presence of the following mineral phases in all coal fly ash, quartz, mullite, magnetite, and hematite (**Figure 2**). The presence of the amorphous phase can be identified as the broad diffraction "hump" in the region between 18° and 32° 2θ [21, 26]. The quantitative analysis results show that mullite, quartz, and amorphous materials comprised 96% of the total mineral composition (**Figure 3**). It is noticeable in **Figure 3** that coal fly ash A has a higher content of the amorphous phase and the lowest quantity of quartz, mullite, hematite, and magnetite phases. Rayalu et al. [22] pointed out that the low levels of mullite promote zeolite synthesis and this is attributed to the fact that mullite is resistant to dissolution during hydrothermal treatment. The presence of the aluminosilicate phases in all the evaluated coal fly ashes qualifies their potential for the conversion to zeolite Na-P1.

3.2. Optimization of zeolite synthesis conditions

The optimization of the zeolite synthesis conditions was carried out using coal fly ash A. These optimized conditions were then applied for zeolite synthesis using coal fly ashes B and C and the properties of the zeolitic material compared to that resulting from coal fly ash A. The optimized conditions were the hydrothermal treatment temperature, time, and amount of water during the hydrothermal treatment process. The results are presented in terms of the chemical and mineralogical analysis of the zeolitic products, morphology, CEC, surface area, and pore volume.

3.2.1. Optimization of the hydrothermal treatment temperature and time

XRD was employed as a tool to monitor the evolution of zeolitic phases during the hydrothermal treatment process. Several authors pointed out that the increase of temperature leads to an increase in mullite dissolution, nucleation, and growth in zeolite crystals and an increased in Si dissolution from coal fly ash leading to an increase in Si/Al ratio, which influences the crystallinity of the final zeolitic product [10, 33, 34]. **Figure 4** shows the XRD profile of zeolitic products for a concurrent variation of hydrothermal treatment time and temperature for coal fly ash A feedstock. **Figure 4a** shows the XRD profile of the transformation of different phases of the coal fly ash as a function of time at hydrothermal treatment temperature of 100°C. **Figure 4a** shows the disappearance of the broad hump between 18° and 32° 2θ (**Figure 2**), signifying the dissolution of the amorphous glassy phase during the hydrothermal treatment. The zeolitic phase produced was identified as zeolite Na-P1 ($Na_6Al_6Si_{10}O_{32}$ċ$12H_2O$). Some quartz, mullite, and hematite phases remained undissolved from the matrix even after 48 h hydrothermal treatment time at 100°C. **Figure 4b** shows the XRD profile of the transformation of different phases of the coal fly ash as a function of time at the hydrothermal treatment temperature of 120°C. The hydrothermal treatment at 120°C for various times did not result in pure-phase zeolite. An incomplete dissolution of the coal fly ash matrix was also observed with mullite and quartz being identified. However, a new zeolitic phase was identified as hydroxy-sodalite formed in all treatment times. At an increased hydrothermal treatment temperature of 140°C (**Figure 4c**), an increased dissolution of the coal fly ash aluminosilicate matrix was observed with increased hydrothermal treatment time, leading to almost complete dissolution at 48 h. The hematite mineral phase was completely dissolved. At 160°C hydrothermal treatment temperature, the mullite and quartz phases were gradually dissolved as the treatment time increased with complete dissolution at 48 h. However, another zeolitic phase was observed (hydroxyl-sodalite, whose peaks increased in intensity with increasing time; **Figure 4d**). The optimization of the hydrothermal treatment time and temperature identified

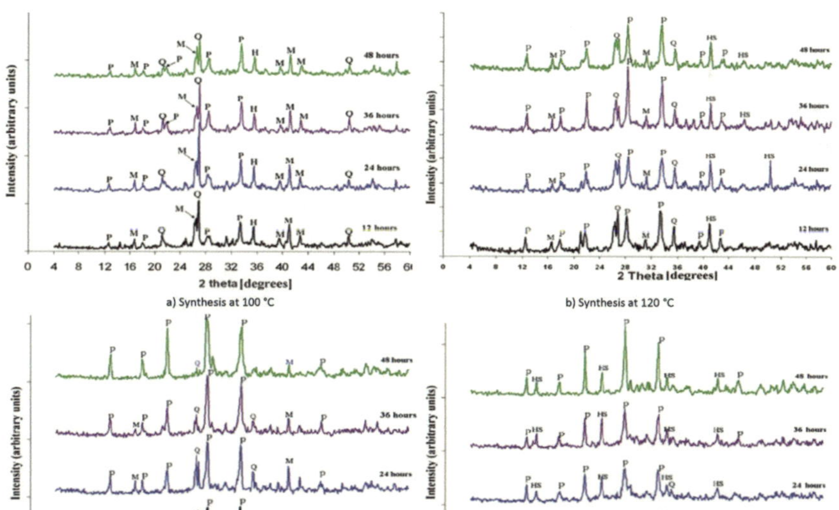

Figure 4. XRD profile of zeolitic products on the concurrent variation of hydrothermal treatment time (12–48 h) and temperature (100–160°C) for coal fly ash A (P = zeolite Na-P1, Q = quartz, M = mullite, H = hematite, HS = hydroxy-sodalite). Reproduced with permission from Taylor & Francis Group, LLC, the copyright owners.

140°C temperature and 48 h as the optimum conditions for the synthesis of pure-phase zeolite Na-P1. However, traces of mullite and quartz were still observed in the almost pure phase obtained.

3.2.2. *Effect of variation of water during hydrothermal treatment process*

During the aging process, the dissolution of the aluminosilicate matrix of the coal fly ash occurs, releasing Si and Al, which are the main building blocks of the zeolite. Depending on the conditions employed, the saturation of the ionic precursors can occur, hindering more dissolution. The addition of water will tend to dilute the concentration of the ionic precursors leading to the further dissolution of the coal fly ash matrix. **Figure 5** shows the XRD profile of zeolitic products obtained on varying the water added during the hydrothermal treatment at 140°C and 48 h treatment time.

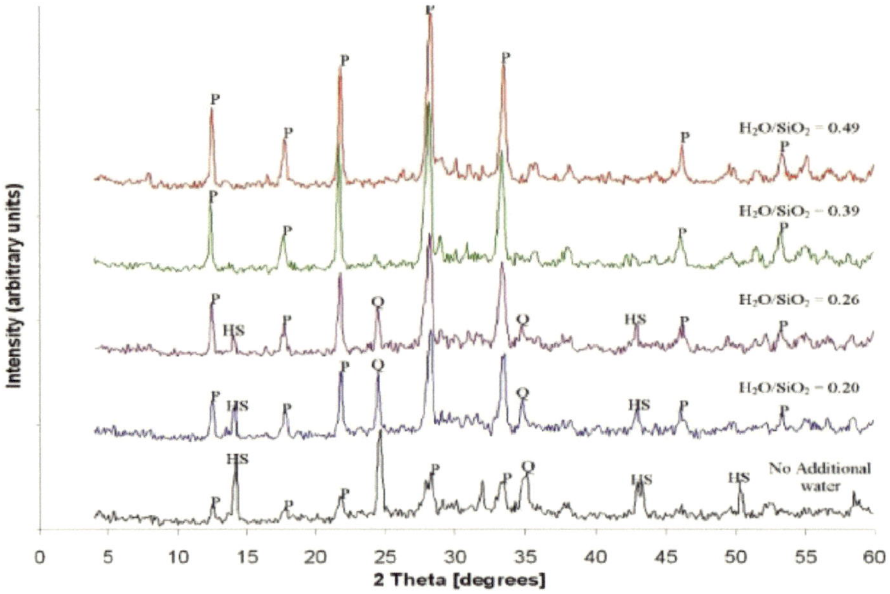

Figure 5. XRD profile of zeolitic products synthesized at 140°C for 48 h with variation of water during the hydrothermal treatment (P = zeolite Na-P1, HS = hydroxy-sodalite, Q = quartz). Reproduced with permission from Taylor & Francis Group, LLC, the copyright owners.

The additional water was added after the aging process, as equal amounts of water (100 mL) had been used during the aging process. It is observed that the addition of more water after aging enabled the formation of relatively pure phase of zeolite Na-P1. A balance has to be struck on the amount of water to be added, as too much water can also change the degree of supersaturation, which can slow the crystallization kinetics in addition to increasing the cost of zeolite synthesis due to the need to manage the disposal of large amounts of effluent generated.

3.2.3. Relative percent crystallinity of the zeolitic products

The relative percent crystallinity of the zeolitic products was done by summing and normalizing the peak heights of five major peaks (2θ=12°, 23°, 28°, 33°, and 46°; **Figure 4c**). The normalized peaks of zeolite Na-P1 synthesized at 140°C for 48 h was assumed to be 100% crystalline for the purpose of comparing to the spectra of the zeolites synthesized at 12, 24, and 36 h. A comparison of the percent crystallinity (peak heights) of the zeolite Na-P1 with quartz peaks was observed to be inversely proportional (**Figure 6**). This suggested the dissolution of quartz phase and the release of Si, which was then made available for the zeolitization.

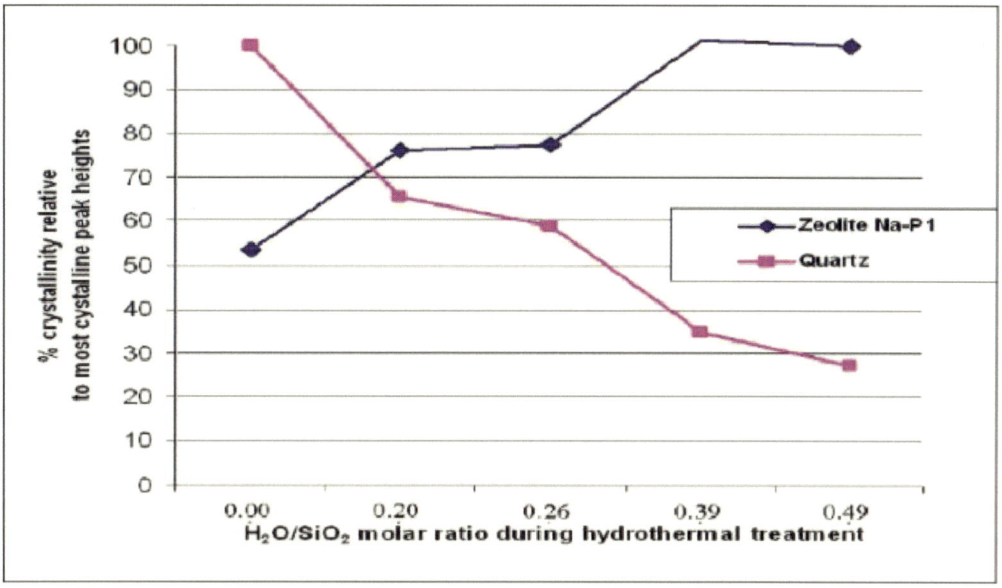

Figure 6. Comparison of the relative percent crystallinity of quartz and zeolite Na-P1 synthesized at 140°C (48 h) by varying the H_2O/SiO_2 molar ratio during the hydrothermal treatment.

3.3. Synthesis of zeolite Na-P1 using other coal fly ashes at optimized conditions

To test the success of the optimization process, the synthesis of zeolite Na-P1 from different coal fly ash feedstock was attempted at the optimized conditions. The XRD and XRF characterization of these coal fly ashes (coal fly ashes B and C) were presented previously (**Figures 2** and **3**; **Table 1**). **Figure 7** presents a comparison of the XRD profile of zeolite Na-P1 synthesized from coal fly ashes A, B, and C at the previously optimized conditions (140°C hydrothermal temperature and 48 h treatment time). **Figure 7** indicates that the zeolite Na-P1 produced with coal fly ash B was similar to that of coal fly ash A. However, coal fly ash C showed an incomplete dissolution of mullite and quartz. This could be attributed to the difference in chemical and mineralogical composition of the coal fly ash C compared to coal fly ashes A and C (**Figures 2** and **3**; **Table 1**). This indicates that the conditions employed were not optimum for the dissolution of quartz and mullite in coal fly ash C. Another observable fact about the incomplete zeolitization of coal fly ash C could be due to the low CaO, which is known to contribute to the alkalinity of the solution. Coal fly ash C had the lowest CaO content (**Table 1**).

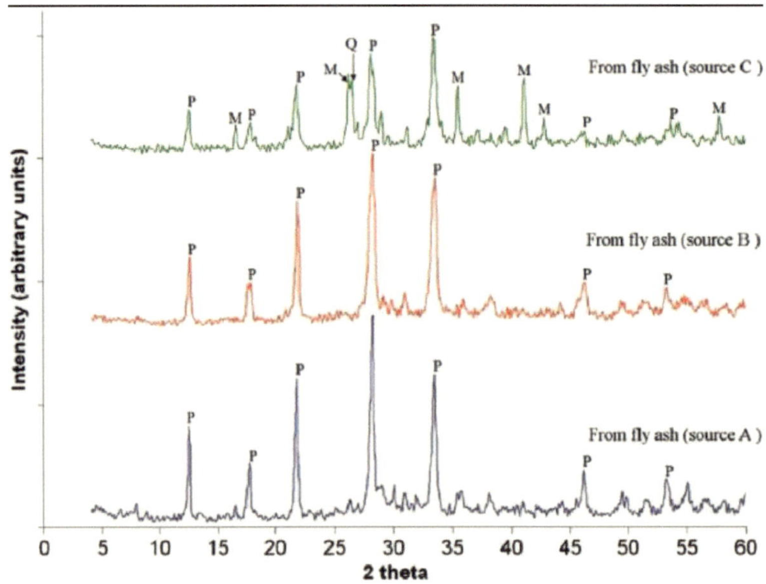

Figure 7. XRD profile of zeolites synthesized from coal fly ashes A, B, and C at 140°C and 48 h treatment time. Reproduced with permission from Taylor & Francis Group, LLC, the copyright owners.

3.4. Morphological evolution of the coal fly ashes during the synthesis process

The morphological evolution of the coal fly ash feedstock after aging and hydrothermal treatment is depicted in **Figure 8**. The coal fly ash has a smooth spherical appearance and typically consists of microspheres. Gitari et al. [15] have previously reported the smooth spherical appearance of South African coal fly ash. The coal fly ash particles are observed to be smooth spheres, as the glassy phase covers the particles. The coal fly ash after the aging process appears to show a partial disintegration of the spheres, indicating the conditions employed during aging are not sufficient to completely break the matrix. This could also point to the resistance of mullite and quartz to dissolution. However, after hydrothermal treatment, a transformation of the disintegrating microspheres was observed, with granular crystalline particles being observed on the surface of the spheres. These were identified by XRD to be the zeolitic phases. Walek et al. [24] reported that the hydrothermal crystallization process begins on the surface of the undissolved or partially dissolved coal fly ash particle. These granular particles represented the initial zeolite formation process. A further illustration of the evolution of the crystalline zeolitic phases is depicted during the variation of water during the hydrothermal treatment at 140°C temperature for 48 h (**Figure 9**). Crystalline granules are clearly evident on the surface of the microspheres, which increases as the amount of water is increased. The crystallinity is also evident from the increasing agglomeration of the crystalline granules. The formation of the well-ordered crystalline zeolitic phase formed during aging conditions of 47°C for 48 h and hydrothermal treatment for 48 h at 140°C with H_2O/SiO_2 ratio of 0.49 was confirmed by use of HRTEM (**Figure 10**).

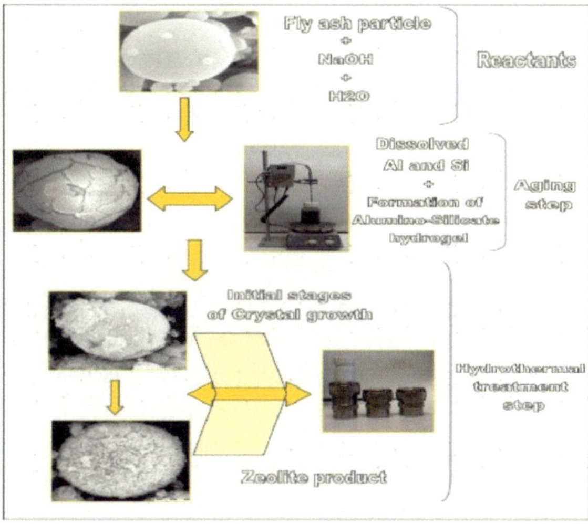

Figure 8. Morphological illustration of the reaction mechanism for the formation of zeolite phases.

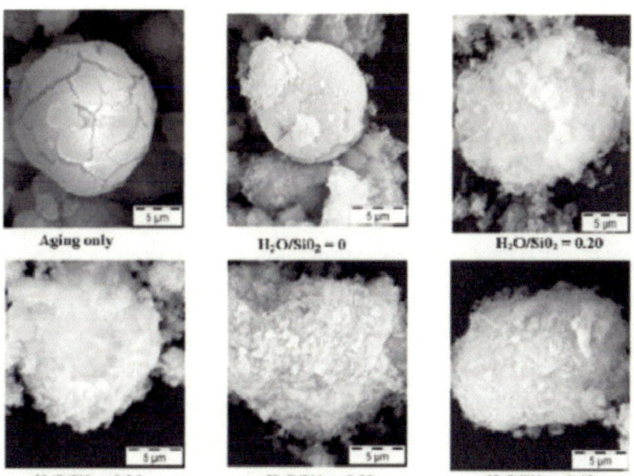

Figure 9. Morphological evolution of the zeolitic phases with variation of water content during the hydrothermal treatment.

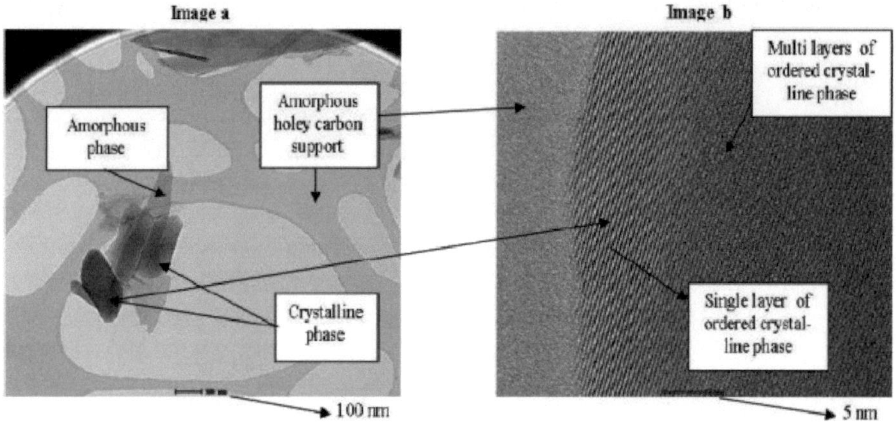

Figure 10. HRTEM micrograph showing the well-ordered crystalline zeolitic phase obtained from coal fly ash A feed-stock.

In conclusion, the observations on the morphological transformation/evolution of the coal fly ash indicate that the glassy phase was dissolved in alkaline media and subsequently transformed into zeolite crystals, which were deposited on the surface of the disintegrating microsphere. The zeolite formation mechanism can therefore be described as (1) the dissolution of the SiO_2 and Al_2O_3 from the glassy phase/mullite/quartz in alkaline media, (2) the formation of aluminosilicate gel as zeolite precursor from SiO_4 and AlO_4 species, and (3) the crystallization of the zeolite phase at supersaturation of the alkaline media. The addition of water during hydrothermal treatment lowers the supersaturation and allows more dissolution of the glassy phase/mullite/quartz and the release of more SiO_4 and AlO_4 species leading to increased formation of the zeolite phase. The disintegration of the fly ash microsphere is predicted to continue with increased dissolution of the resistant mullite and quartz phases, as more hydroxyl ions migrate through the microsphere with continued hydrothermal treatment and subsequent formation of crystalline zeolitic phases (**Figure 9**).

3.5. CEC and surface area

The CEC of the zeolitic products was determined to provide further evidence to the purity of the phases. An increase in CEC was observed in the zeolitic products compared to the feedstock coal fly ash. Coal fly ash had a CEC of 0.48 mEq/g, zeolite product at 100°C/48 h of 2.98 mEq/g, and zeolite product at 140°C/48 h of 3.91 to 4.11 mEq/g. These results correlate well with the increasing crystallinity observed in the optimization stages (**Figures 4** and **9**). The results also compare closely to the CEC value of commercial zeolite Na-P1 (5 mEq/g). The addition of water during the hydrothermal treatment enhanced the CEC of the zeolitic products (zeolitic product obtained at 140°C/48 h with a H_2O/SiO_2 molar ratio of 0.49 was 4.10 mEq/g). This again confirms the high crystallinity and purity of the zeolitic phase obtained at this synthesis conditions. The BET analysis of surface area showed that the surface area of the material obtained after aging was low (30.5693 m^2/g), but this was observed to increase significantly on hydrothermal treatment at 24 h/140°C (58.6358 m^2/g)

and 48 h/140°C (67.6329 m^2/g), which again confirms the high-purity and crystallinity of the zeolitic phase.

4. Conclusions

South African coal fly ashes have been proven to be suitable feedstock for hydrothermal conversion into high-purity zeolite Na-P1. The compositional and mineralogical analysis of the South African coal fly ashes identified mineral phases quartz, mullite, and amorphous glassy phase, which acts as crucial ingredients for the zeolite synthesis process. The optimization of the hydrothermal treatment process identified the conditions for the synthesis of high-purity zeolite Na-P1 as molar regime of 1 SiO_2:0.36 Al_2O_3:0.59 NaOH:0.49 H_2O and synthesis conditions of aging slurry at 47°C for 48 h. This was followed by hydrothermal treatment at 48 h and 140°C. The addition of water after the aging step expressed as H_2O/SiO_2 molar ratio was observed to play an important role in the hydrothermal conversion of coal fly ash into high-purity zeolites. The results show that a simple adjustment of reactant composition and presynthesis or synthesis parameters leads to almost complete dissolution of coal fly ash matrix and the conversion of the Si/Al ionic precursors into high-purity zeolite Na-P1. The high purity and CEC (4.11 mEq/g) imply that the zeolites from the class F South African coal fly ash can meet high-end applications such as wastewater treatment or application in detergents.

Acknowledgements

The authors would like to acknowledge the Water Research Commission (WRC), National Research Foundation (NRF), and Coaltech 2020 for funding this research work.

Author details

Mugera W. Gitari[1*], Leslie F. Petrik[2] and Nicholas M. Musyoka[3]

*Address all correspondence to: mugera.gitari@univen.ac.za

1 Environmental Remediation and Water Pollution Chemistry Group, School of Environmental Sciences, University of Venda, Thohoyandou, South Africa

2 Environmental and Nano Science Research Group, Department of Chemistry, University of the Western Cape, Cape Town, South Africa

3 Materials Science and Manufacturing, Council for Scientific and Industrial Research, Pretoria, South Africa

References

[1] Petrik LF, White R, Klink M, Somerset V, Key D, Iwuoha E, Burger C, Fey MV. Utilization of fly ash for acid mine drainage remediation. Pretoria, South Africa: Water Research Commission; 2005. 174 pp. ISBN: 1-77005-316-6

[2] ESKOM Annual Report. 2010. Available at: www.eskom.co.za

[3] ESKOM Annual Report. 2001. Available at: www.eskom.co.za

[4] Gitari WM, Leslie FP, Key DL, Okujeni C. Partitioning of major and trace inorganic contaminants in fly ash acid mine drainage derived solid residues. International Journal Environmental Science Technology. 2010; 7: 519–534. ISSN: 1735-1472

[5] Gitari WM, Fatoba OO, Leslie FP, Vadapalli VRK. Leaching characteristics of selected South African fly ashes: Effect of pH on the release of major and trace species. Journal of Environmental Science and Health Part A. 2009; 44: 206–220. DOI: 10.1080/10934520802539897

[6] Molina A, Poole C. A comparative study using two methods to produce zeolites from fly ash. Minerals Engineering. 2004; 17: 167–173. DOI: 10.1016/j.mineng.2003.10.025

[7] Ye Y, Zeng X, Qian W, Mingwen W. Synthesis of pure zeolites from supersaturated silicon and aluminium alkali extracts from fused fly ash. Fuel. 2008; 87: 1880–1886.

[8] Holler H, Wirsching U. Zeolite formation from fly ash. Fortschritte der Mineralogie. 1985; 63: 21–43.

[9] Shigemoto N, Hayashi H, Miyama K. Selective formation of Na-X zeolite from coal fly ash by fusion with sodium hydroxide prior to hydrothermal reaction. Journal of Materials Science. 1993; 17: 4781–4786. DOI: 10.1007/BF00414272

[10] Querol X, Plana F, Alastuey A, López-Soler A. Synthesis of Na-zeolites from fly ash. Fuel. 1997; 76: 793–799. DOI: 10.1016/S0016-2361(96)00188-3

[11] Hollman GG, Steenbruggen G, Janssen-Jurkovicová M. A two-step process for the synthesis of zeolites from fly ash. Fuel. 1999; 78: 1225–1230. DOI: 10.1016/S0016-2361(99)00030-7.

[12] Woolard CD, Petrus K, Van der Horst M. The use of modified fly ash as a sorbent for lead. Water SA. 2000; 26: 531–536. ISSN: 0378-4738

[13] Querol X, Umana F, Plana F, Alastuey A, López-soler A, Medinaceli A, Valero A, Domingo MJ, Garcia-rojo E. Synthesis of zeolites from fly ash at pilot scale. Example of potential applications. Fuel. 2001; 80: 857–865. DOI: 10.1016/S0016-2361(00)00156-3

[14] Somerset VS, Petrik LF, White RA, Klink MJ, Key D, Iwuoha E. The use of X-ray fluorescence (XRF) analysis in predicting the alkaline hydrothermal conversion of fly

ash precipitates into zeolites. Talanta. 2004; 64: 109–114. DOI: 10.1016/j.talanta. 2003.10.059

[15] Gitari MW, Petrik LF, Etchebes O, Key DL, Iwuoha E, Okujeni C. Treatment of acid mine drainage with fly ash, removal of major contaminants and trace elements. Journal Environmental Science Health Part-A Toxicology Hazard Substances Environmental Engineering. 2006; 41: 1729–1747. DOI: 10.1080/10934520600754425

[16] Derkowski A, Franus W, Beran E, Czímerová A. Properties and potential applications of zeolitic materials produced from fly ash using simple method of synthesis. Powder Technology. 2006; 166: 47–54. DOI: 10.1016/j.powtec.2006.05.004

[17] Flanigien EM. Zeolite and molecular sieves: A historical perspective. In: Van Bekkum H, Flanigen EM, Jansen JC, editors. Introduction to Zeolite Science and Practice. 2nd ed. Amsterdam, Netherlands: Elsevier Science; 1991. pp. 13–33. ISBN: 978-0-444-82421-9

[18] Singer A, Berkgaut V. Cation exchange properties of hydrothermally treated coal fly ash. Environmental Science and Technology. 1995; 29: 1748–1753. DOI: 10.1021/es00007a009

[19] Querol X, Moreno N, Umaña JC, Alastuey A, Hernández E, López-Soler A, Plana F. Synthesis of zeolites from fly ash: An overview. International Journal of Coal Geology. 2002; 50: 413–423. DOI: 10.1016/S0166-5162(02)00124-6

[20] Inada M, Tsujimoto H, Eguchi Y, Enomoto N, Hojo J. Microwave assisted zeolite synthesis from fly ash in hydrothermal process. Fuel. 2005; 84: 1482–1486. DOI: 10.1016/j.fuel.2005.02.002

[21] Murayama N, Yamamoto H, Shibata J. Mechanism of zeolite synthesis from fly ash by alkali hydrothermal reaction. International Journal Mineral Processing. 2002; 64: 1–17. DOI: 10.1016/S0301-7516(01)00046-1

[22] Rayalu SS, Udhoji JS, Munshi KN, Hasan MZ. Highly crystalline zeolite—A from flyash of bituminous and lignite coal combustion. Journal of Hazardous Materials. 2001; 88: 107–121. DOI: 10.1016/S0304-3894(01)00296-5

[23] Chareonpanich M, Namto T, Kongkachuichay P, Limtrakul J. Synthesis of ZSM-5 zeolite from lignite fly ash and rice husk ash. Fuel Processing Technology. 2004; 85: 1623–1634. DOI: 10.1016/j.fuproc.2003.10.026

[24] Walek TT, Saito F, Zhang Q. The effects of low solid/liquid ratio on hydrothermal synthesis of zeolites from fly ash. Fuel. 2008; 87: 3194–3199. DOI: 10.1016/j.fuel. 2008.06.006

[25] Zholobenko VL, Dwyer J, Zhang R, Chapple AP, Rhodes NP, Stuart JA. Structural transitions in zeolite P. An in situ FTIR study. Journal of Chemical Society. 1998; 94: 1779–1781. DOI: 10.1039/A800060C

[26] Hui KS, Chao CYH. Effects of step-change of synthesis temperature on synthesis of zeolie 4A from fly ash. Microporous Mesoporous Materials. 2005; 88: 145–151. DOI: 10.1016/j.micromeso.2005.09.005

[27] Petrik LF, Hendricks N, Ellendt A, Burgers C. Toxic element removal from water using zeolite adsorbents made from solid waste residues. Pretoria, South Africa: Water Research Commission; 2007. 133 pp. ISBN: 1770054642, 9781770054646

[28] Szostak R. Molecular sieves, principles of synthesis and identification. In: Van Nostrand Reinhold Catalysis Series. The Netherlands: Springer; 1989. DOI: 10.1007/978-94-010-9529-7

[29] Casci JL. Zeolite molecular sieve: Preparation and scale up. Microporous and Mesoporous Material. 2005; 82: 217–226. DOI: 10.1016/j.micromeso.2005.01.035

[30] Gitari MW, Petrik LF, Reynolds K. Chemical weathering in a hypersaline effluent irrigated dry ash dump: An insight from physicochemical and mineralogical analysis of drilled cores. Energy Science and Technology. 2011; 2: 43–55. DOI: 10.3968/j.est.1923847920110202.110

[31] Amrhein C, Gaghnia GH, Kim TS, Mosher PA, Gagajena RC, Amanios T, Torre IDL. Synthesis and properties of zeolites from coal fly ash. Environmental Science and Technology. 1996; 30: 735–742. DOI: 10.1021/es940482c

[32] Gitari M, Petrik LF, Key D, Etchebers O, Okujeni C. Mineralogy and trace element partitioning in coal fly ash/acid mine drainage co-disposed solid residues. In: Proceedings of the World of Coal Ash Association Conference, Lexington, KY, USA; 11–15 April 2005. pp. 1–21.

[33] Catalfamo P, Pasquale SD, Corigliano F, Mavilia L. Influence of the calcium content on the coal fly ash features in some innovative applications. Resources, Conservation and Recycling. 1997; 20: 119–125. DOI: 10.1016/S0921-3449(97)00013-X

[34] Feijen EJP, Martens JA, Jacobs PA. Zeolites and their mechanism of synthesis. In: Proceedings of the 10th International Zeolite Conference, Garmisch-Partenkirchen, Germany; 17–22 July 1994. pp. 3–21.

Chapter 3

Synthesis of Zeolite from Fly Ash and their Use as Soil Amendment

Maria Harja, Sorin Mihai Cimpeanu,
Marcel Dirja and Daniel Bucur

Additional information is available at the end of the chapter

http://dx.doi.org/10.5772/64126

Abstract

The amendments may be natural or synthetic and by their origin organic or inorganic. Among amendments, the inorganic ones are recommended, such as gypsum, lime, zeolites and altered rocks, including sand, ceramic, perlite, ash, slag, zeolite, pyrites, dolomite, calcined clay, etc. Fly ash-based zeolites can be used as amendments to improve the use of the soil for agricultural purposes. Generally, there are different methods for the synthesis of new materials using solid waste – fly ashes. Methods are known to be composed of a single stage or two stages. The first method – the conventional – is direct hydrothermal conversion of the mixture of ash and alkaline solution (NaOH or KOH), but only 50% of ashes can be converted into zeolite. The second method consists in mixing the ash with KOH and fusion at elevated temperature. The method leads to an advanced conversion – the type of zeolite depending on treatment conditions. Other studies recommend the conversion of ash using microwave ovens or ultrasound bath; in this case, the conversion time is reduced to 1–2 hours. From these methods, the direct method was proposed to be used in synthesizing zeolites for agriculture, with a part in controlling release of fertilizers and as a soil amendment. The synthesized materials were characterized with respect to microstructure (electronic microscopy SEM) and chemical and mineralogical composition (EDAX, X-ray diffraction). By analyzing the obtained data, one may observe the destruction of the ash network and crystallization of the zeolitic phase, especially in the case of treatment by diffusion. The XRD analysis confirms the presence of zeolite in materials obtained. Good results were also obtained in the case of using ultrasound treatment for zeolite synthesis; also, the time of treatment significantly decreased in this case.

Keywords: amendment, fly ash, heavy metals, characterization, synthesis, soil, zeolite

1. Introduction

Soil is the most important component of the surface crust along with water and air of which it is composed of the environment. The mineral composition of soils is very diverse, usually including minerals from the silicate group, oxides, and hydroxides, and minerals ordered mainly on the grain size criteria [1]. The usual minerals are quartz, feldspar, mica and calcite, mullite, kaolinite, montmorillonite, illite, vermiculite, and chlorite. Organic and mineral constituents of soils are represented by organic complexes of metal ions (Cu^{2+}, Fe^{2+}, Mn^{2+}) and associations between clay and organic humic and nonhumic substances.

In the last decade, there has been a decrease in the areas with agricultural uses due to decreased fertility, compaction, acidification, heavy metal pollution, etc. To achieve higher agricultural production, measures must be taken to improve the physical, chemical, and biological properties of soils. Due to the acidity in the soil, nitrification and nitrogen-fixing soil bacteria do not develop; therefore, the content of nitrogen is low [2].

For improving soil characteristics, the research on soil quality being oriented toward finding measures to improve the structure, air, and water regime, restoring the content of organic matter, improving activity of soil microorganisms, knowing the transformation to which applied chemicals are exposed, critical limits of the pollution, reducing the content of heavy metal ions, and evaluating the risks to human health and domestic animals and wildlife.

A soil amendment describes any substances added to a soil to improve or modify its qualities and the conditions including physical, chemical, and biological conditions. An amended soil is a conditioned and a fertilized one. Choosing the amendment depends on what we seek to improve, the degree of soil degradation, its composition, the climate, etc. A suitable amendment to a situation may not be useful in other conditions [3].

The main objective of the amendments is to improve the use of the soil for agricultural purposes. The amendments may be natural or synthetic and by their origin organic or inorganic. Among amendments the inorganic ones are recommended, such as gypsum, lime, zeolites, altered rocks, ceramic, perlite, ash, slag, zeolite, pyrites, dolomite, calcined clay, etc. They contain a variable amount of nutrients, which can affect the pH.

Among inorganic substances used, the study focuses on thermal power station ash and its use in the synthesis of zeolites.

Among industrial wastes, ash plays a leading role because of the large volume and the consequences, due to a growing accumulation. Annually about 600 million tons of ash are produced, 75–80% of which is the one resulted from thermal plants, being considered the fifth as a source of prime mover [4].

The ash is used in agriculture since it contains nearly all of micro- and macro-elements, except for the organic carbon and nitrogen [5]. Also, the ash can contain moderate amounts of heavy metals, affecting groundwater and may be retained by plants. This material is used as the filling material, the recovery of acid soils, depending on the pH.

Zeolites are crystalline form of alkaline aluminosilicates, with a stable structure. This structure containing cavities has the ability to retain water and negative charge capably to change heavy metals [6]. This negative charge results in the ability of the zeolites to change plant nutrients and eliminate them from the structure. This ability to exchange cations allows nutrients such as nitrogen, calcium, iron, magnesium, potassium, and water to be available in the plant and to be removed at a low speed in time. In this way, the plant can absorb nutrients from the soil whenever it is required. Zeolites improve performance of fertilizers by increasing resistance to wash, immobilization, and gas losses. They increase soil water reserves needed during dry periods and are able to reversibly lose and retain water and to change network's cations without the change in the structure. They are chemically inert, nontoxic to plants, animals, humans, and the environment and regulate the balance of soil through the immobilization of toxic contaminants.

2. Methods and types of amendments used for reduction of heavy metals

The contamination of soil with heavy metals has been occurring from the last century, but its extent has increased in the last years, due to industrial progresses and amplified user of supplies containing these metals. In the category, heavy metals enter a series of chemical elements, with high toxicity to living organisms. The toxic effect occurs above a certain threshold below which some of them (Co, Cd, Cu, Fe, Ni, Zn, Hg, and Pb) may even be essential components of proteins involved in various metabolic pathways. Thus, if the food would be completely devoid of metals, it would appear nutritional deficiencies.

The heavy metals are found in various concentrations in the soil, water, air, foods of vegetable, or animal origin, depending on the various factors that determine their pollution. Air can be a source of contamination representing a way of handling metals and their deposition on soil and plants (e.g., lead emissions from cars).

Accumulation of heavy metals in soils has implications over human health [7]. The heavy metals from different ecosystems would migrate into groundwater or bioaccumulate in herbs and get into the food chain [8]. In general, heavy metals may affect the soil ecosystem, not only agricultural produce and groundwater quality but also the human health [9].

Heavy metal contamination of the air is the result of numerous anthropogenic activities: combustion of coal, petroleum, nonferrous metal production, production of iron, steel and cement production, installations for waste gas treatment and incineration of waste accumulation, etc. The sources of metals in the soil can be use of fertilizers, pesticides containing metals (fungicides containing mercury, copper, arsenic, zinc, etc.). Of course, depending on the type of soil and geographic location, it can contain high amounts of heavy metals or they can be deficient.

2.1. Methods used for reduction of heavy metals

Many methods such as solidification, cementation, phytoremediation, earth-swap, and soil flushing can be used to reduce the heavy metal from contaminated soils. Among them, fixation of heavy metal using amendments is a promising technology for removing heavy metals from contaminated soils. The application of soil amendments modified the properties of the contaminating heavy metals that allow their immobilization [10–12].

The fixation of heavy metals in soils had been greatly carried out with different amendments, such as zeolite, phosphates, cements, limes, and carbonates, or in the last years, with synthetic amendments: zeolites [13, 14], geopolymers, and phosphate [15–17]. In recent years, there were a lot of studies regarding amendments including by-products, waste, and low-cost synthetic materials for fixation of heavy metals. Low-cost amendments are required for the enhancement of fixing efficiency.

The fixation treatment is based on the reduction of metal mobility and availability as a result of applying amendments. In this way, the unfavorable effects of heavy metals to environmental receptors are reduced [18]. The mechanism of heavy metals removal is based on adsorption, precipitation, cation exchange, and complexation.

2.2. Types of amendments

The fixation efficiency depends on the quantity of the amendments and on the properties of these and soils. The chemical and physical properties are also the influencing factors in the fixation process. The cation mobility is the greatest property that can determine their environmental impact. The heavy metal ions have the individual mobilities and are difficult to find the amendments for reducing the bioavailability.

Among the numerous treatment compounds, some are recommendable for several heavy metals: limes, phosphates, organic matter, industrial by-products, and low-cost zeolites.

Materials	Heavy metals	References
Fly ash	Cd, Pb, Cu, Zn, Cr	[22, 23]
Lime	Cd, Cu, Ni, Pb, Zn, Cr	[18, 19]
Phosphate salt	Pb, Zn, Cd, Cu	[3, 16]
Hydroxyapatite	Zn, Pb, Cu, Cd	[15]
Slag	Cd, Pb, Zn, Cr	[24]
Ettringite	Cd, Cu, Pb, Zn, Cr	[25]

Table 1. Types of inorganic amendments for metal immobilization [18].

Firstly, additives used for heavy metal immobilization can be classified into inorganic, organic, and both inorganic-organic materials. According to the nature of different materials, there are three resources including natural materials, artificial synthetics, and by/co-products used for fixing process. Natural materials that are available in a large quantity or are waste products from industrial or agricultural sources may have potential as low-cost sorbents. Due to their low cost, after utilization they can be disposed of without expensive regeneration. In addition, inorganic or organic amendments are used to improve soils' substrate characteristics. The types of inorganic amendments are presented in **Table 1**.

The materials including CaO, $Ca(OH)_2$, $CaCO_3$, $CaMgCO_3$, $CaHPO_3$, $Ca(H_2PO_3)_2$, K_2HPO_4, H_3PO_4, and $(NH_4)HPO_4$ are the typical amendments used for the immobilization of heavy metals. On the other hand, zeolites decrease the metal concentration in soil solution, the metal mobility, and leachability and transform heavy metal cations from soluble into insoluble cations [16, 19, 20]. Wang et al. [21] reported that $Ca(H_2PO_4)_2$ could replace $CaHPO_4$ due to lower cost and higher solubility in the field study. The combination of $Ca(H_2PO_4)_2$ and $CaCO_3$ significantly reduces the extractable metal concentration and successfully immobilize heavy metals on the site. Quicklime (CaO) was a more effective additive because it is readily soluble and available for reactions, and its addition to a soil material results in larger pH increases.

Materials	Abbreviations	Chemical form
NaP1	NaP1	$Na_6Al_6Si_{10}O_{32}\cdot 12H_2O$
Sodalite	S	$Na_6(H_2O)_8Si_6Al_6O_{24}$
Analcime	A	$NaAlSi_2O_6\cdot H_2O$
Faujasit	Fau	$Na_2Al_2Si_{3.3}O_{8.8}\cdot 6.7H_2O$
Tobermorite	T	$Ca_5(OH)_2Si_6O_{16}\cdot 4H_2O$
Na-X	Na-X	$Na_{88}(H_2O)_{220}Si_{104}Al_{88}O_{384}$
Na-Y	Na-Y	$Si_{192}O_{384}$
Zeolite A	Zeolitul A	$NaAlSi_{1.1}O_{4.2}\cdot 2.25H_2O$
Hydroxy-sodalite	HS	$Na_{1.08}Al_2Si_{1.68}O_{7.44}\cdot 1.8H_2O$
Zeolite X	X	$NaAlSi_{1.23}O_{4.46}\cdot 3.07H_2O$
Zeolite Y	Y	$NaAlSi_{2.43}O_{6.86}\cdot 4.46H_2O$
Cancrinite	Can	$Na_7Ca_{0.9}(CO_3)_{1.4}(H_2O)_{2.1}\ Si_6Al_6O_{24}$
Gismondine	Gis	$Ca_{3.68}(H_2O)_{17.2}Si_{8.8}Al_{7.2}O_{32}$
Chabazite	Cha	$K_{3.2}Na_{0.75}(H_2O)_{8.4}Si_{8.2}Al_{3.8}O_{24}$
Linde A	Linde A	$Na_{96}(H_2O)_{216}Si_{96}Al_{96}O_{384}$
Clinotobermorite	CT	$Ca_5(Si_6O_{17})\cdot 5H_2O$
Nephelin	Nep	$Na_3H_2OSi_3Al_3O_{12}$
Tobermorite	T	$Ca_5(OH)_2Si_6O_{16}\cdot 4H_20$
Phillipsite	Ph	$K_2Al_2Si_3O_{10}\cdot H_2O$
K-chabazite	K-Cab	$K_2Al_2SiO_6\cdot H_2O$
Linde F	Linde F	$KAlSiO_4\cdot 1.5H_2O$
Kalsilite	K	$KAlSiO_4$

Table 2. Chemical composition of zeolite obtained from fly ash [31, 32].

Zeolites are crystalline aluminosilicate form alkali [26, 27]. They have a stable structure with three-dimensional five silica tetrahedra in a unique symmetrical arrangement [28]. This structure containing cavities has the ability to retain water and negative charge inside the opening cation binds to change [29]. This negative charge results in the ability of zeolites to change plant nutrients and eliminate them from the structure [29].

Zeolites are recognized as "intelligent fertilizers" because of the ion exchange capacity (CEC) and high porosity [30]. The ability to exchange cations allows nutrients such as nitrogen, calcium, iron, magnesium, potassium, and water to be available in the plant and be eliminated slowly [26, 29, 30]. In this way, plants can absorb nutrients from the soil when required. Zeolites improve performance by increasing resistance to wash fertilizers, immobilization, and gas leaks. They increase water reserves in soil needed during dry periods and are able to reversibly lose and retain water and change cations' network without changing the structure. They are chemically inert, nontoxic to plants, animals, humans, and the environment and regulate the balance of soil by immobilization of toxic contaminants to the plant [26].

As a result, the application of the zeolite increases ion-exchange capacity, which allows the retention followed by a controlled release of nutrients, reduces the consumption of fertilizer and water necessary to complete the potassium, increases the availability of the use of phosphorous adsorbed toxic contaminants, acts as a buffering agent pH and alkalinity, reduces the effects, improves the aeration, allows the activity of microorganisms, establishes the alkali metal balances, and is cheap. It can have nutritional value that helps germination, root development, and plant growth and grain [28–30].

Zeolites with chemical composition presented in **Table 2** are used, due to the special properties such as adsorption, ion exchange, molecular sieve, hydration, dehydration, bulk density, porosity, and structure [27, 28], to interact with other minerals from the soil, resulting in improved soil structure [26].

3. The methods to obtain zeolites from fly ash

One possibility to recover the ash is the conversion in zeolites. The zeolites prepared by alkali activation of fly ash are aluminum-silicates and include three classes of inorganic polymers, depending on the ratio of silica/alumina: (—Si—O—Al—O—), poly-sialic – SiO_2/Al_2O_3 = 2; (—Si—O—Al—O—Si—O—), poly-sialic siloxo – SiO_2/Al_2O_3 = 4; and (—Si—O—Al—O—Si—O—Si—O—), poly-sialic disiloxo – SiO_2/Al_2O_3 = 6. The composition of the mixture and the reaction conditions such as SiO_2/Al_2O_3 ratio, hydroxide concentration, activation temperature, curing time, ratio of the solid/liquid, and pH significantly affect the formation and properties of zeolites. Limits of variation in these parameters are broad enough: ratio SiO_2/Al_2O_3 2–6, activation temperature 80–150°C, concentration of (Na or K) hydroxide 1–5 M, curing time of 4–48 hours, and the ratio of the solid/liquid 1/1–1/4 [33, 34].

Generally, there are two main methods for the synthesis of new materials using solid waste – fly ashes. Methods are known to be composed of a single stage or two stages.

	1–3 mL/g	
1.0 M NaOH	90°C	Zeolite A
	200°C	NaP1, Her (8 activation hours)
2.0–3.0 M NaOH	150°C	NaP1, Fau
	200°C	NaP1, Her
5.0 M NaOH	150–200°C	NaP1, Her
	200°C	Her, A, HS, HC
2.0 M KOH	150–200°C	KM
5.0 M KOH	150°C	KM, Cha, Linde F
	3–6 mL/g	
0.5 M NaOH	150°C	NaP1(48, 72 activation hours)
1.0 M NaOH	120°C	NaP1, A (6 activation hours)
1.0–3.0 M NaOH	100°C	S, Gis (24, 48, 120, 288 ore)
	10–18 mL/g	
0.5 M NaOH	150°C	NaP1, HS, A (48, 96 activation hours)
		NaP1, T (72 activation hours)
0.5–2.0 M NaOH	95°C, 120°C	NaP1 (24 activation hours)
0.5–2.0 M NaOH	150°C	NaP1, HS, T (9, 24, 48 activation hours)
0.5–2.0 M NaOH	150°C	NaP1, HS, T, A (24, 72 activation hours)
2.0 M NaOH	100°C	4A
0.5–3.0 M NaOH	90–175°C	NaP1
	175–225°C	A, HS, T, Nep
2.0–3.0 M	90°C	Zeolite A
	150°C	NaP1, Fau
	200°C	NaP1, Her
3.0–5.0 M NaOH	150–200°C	HS, HC, Tob
0.5–1.0 M KOH	150–200°C	KM, T
3.0 M KOH	<150°C	Linde F, T
5.0 M KOH	<150°C	Linde F, T

Table 3. Zeolites from fly ash as a function of hydroxide (NaOH/KOH) concentration, temperature and L/s ratio.

In 1985, Holler and Wirsching synthesized, for the first time, zeolites from fly ash by direct activation method [35]. As a result of studies carried out by Holler and Wirsching, several researchers have proposed various methods to obtain new materials based on fly ash in articles and patents:

- direct activation method (hydrothermal method): [36–44];
- fusion method followed by direct activation method: [45–49];
- ultrasound method: [13, 50, 51];
- microwave followed by direct activation method.

Table 3 presents the main types of new materials, which can be obtained as a function of zeolitization conditions.

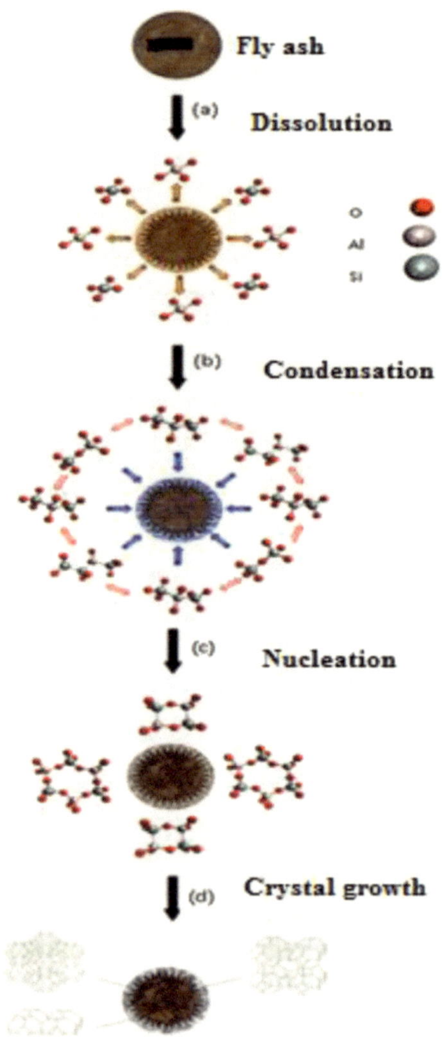

Figure 1. The mechanism of zeolitization of fly ash.

The first method – the conventional one, it is direct hydrothermal conversion of the mixture of ash and alkaline solution. Only 50% of ashes can be converted into P-zeolite or hydroxy sodalite. The second method consists in mixing the ash with alkaline reagents and fusion at elevated temperature (typically 550°C) to convert the ash into a soluble potassium silicate and potassium aluminate, followed by dissolution in an alkaline solution. Fusion method is performed with KOH and NaOH. The method leads to an advanced conversion, the type of zeolite depending on treatment conditions. The third method consists in the extraction of silicon from ashes in the first stage, after that solution was added in a well-established proportion, which leads to a very pure zeolite. Also, experimental studies recommend the conversion of ash using microwave ovens; in this case, the conversion time is reduced from 24–48 hours to 30 minutes. To use this method, extensive research is required because not every type of ash can be changed. Even if the direct method was considered, it is disappointing regarding the zeolite purity and simplicity, and synthesis in a single step was proposed to be used in synthesizing zeolites for agriculture, with a part in controlling release of fertilizers and as a soil amendment.

The main limitation of the synthesis process of zeolites is the temperature (125–200°C) to dissolve Si and Al from the ash particles. If the temperature is reduced, then the yield of the synthesis is considerably reduced, and it is necessary to have a greater reaction time. However, KM, herschelite, K-chabazite, and Linde F can be obtained in the range of 125–200°C.

Fly ash mainly consists of amorphous and subordinate crystalline phases such as quartz, mullite, and hematite. The amorphous material represents the most reactive phase during the zeolitization process. If there is a big content of vitreous phase, the activation time is reduced, and the alkaline hydroxide solution is less concentrated.

After treatment, the samples were allowed to cool for 20 hours at room temperature $T = 20°C$, filtered, washed, and dried.

Murayama et al. [52] and Bukhari et al. [53] showed that the conversion mechanism implicates the steps: dissolution, crystallization, and/or condensation (**Figure 1**).

Either method is used; the most important parameters in the modifying process are activation time; concentration of the solutions used for modification; temperature; and ratio of ash-solution.

3.1. Direct activation method (hydrothermal method)

The experimental results demonstrated that the fly ash consists predominantly of quartz (SiO_2), mullite ($2SiO_2{\cdot}3Al_2O_3$), hematite (Fe_2O_3), and magnetite (Fe_3O_4) can be modified by direct activation method. The method consists in conversion of fly ash into zeolite by treatment of these with alkaline solution, such as NaOH or KOH for extraction of the components (aluminum and silica), and heat treatment for growth of zeolite crystals.

For obtaining one pre-established type of zeolite, it can be possible to insert with aluminum or silica compounds. By the direct activation method, materials were obtained as zeolite X [45, 54, 55], zeolite A, analcime, chabazite, or hydroxy-sodalite as a function of temperature, pressure, Si/Al ratio, alkaline solution, reaction time, pH, etc. [37, 56, 57].

Scheme for the preparation of new materials by direct activation method is shown in **Figure 2**.

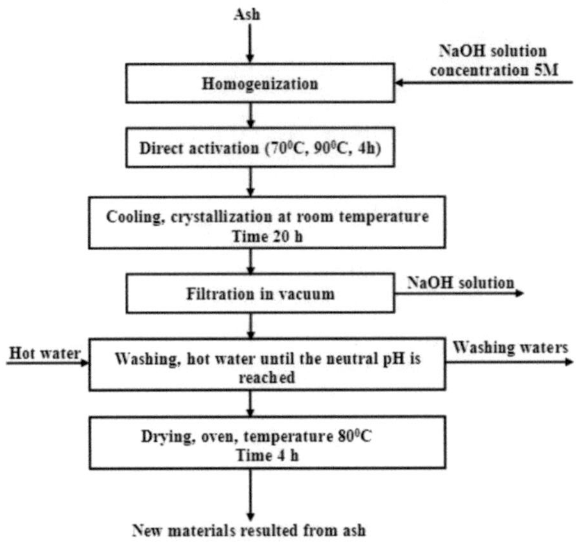

Figure 2. Modification scheme of ash through direct activation.

Zeolite	References
NaP_1	[33, 36–38, 45, 58, 59]
NaA, S	[45, 60]
Na-Cha	[33, 61]
Na-X	[41]
NaP1, T, HC, Her, HS	[58]
NaP_1, A, P, Nep	[36]
NaP_1, HS	[37]
NaP1, HS, T, A	[38]
NaP_1, HS, A	[38, 62]
NaP_1, F	[63]
Na-A	[40]
Gis, S	[47]
Y	[33]
A	[64]
ZSM-5	[65]
HS, NaA, Linde F	[43]
HS, NaP1, X	[66]
Na-HEU	[42]

Table 4. Zeolites obtained by direct activation method.

Direct activation method comprises one step in seeking to use the content of silica in ash to produce a new material. Treatment takes place in an autoclave at temperatures between 80 and 200°C with agents brought to KOH and NaOH for 3–96 hours contact solution concentration: 0.5–5 M and the ratio L/s 1–3 mL/g and 10–18 mL/g (**Table 4**).

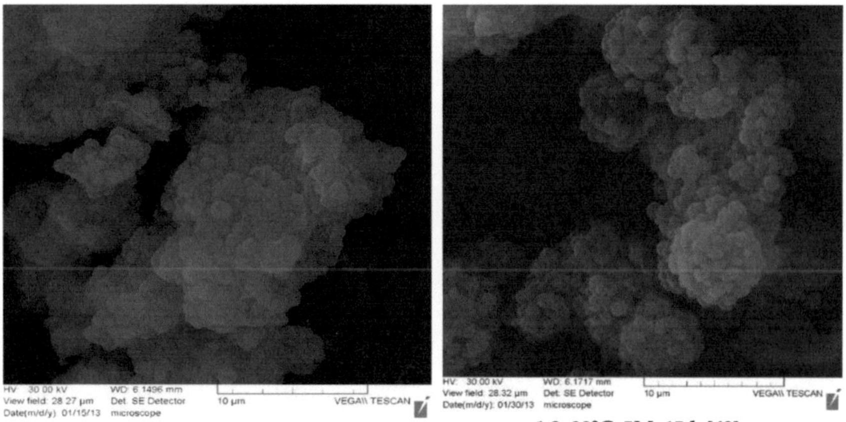

1:5, 90°C, 5M, 4 h, [67] **1:3, 90°C, 5M, 15 h** [68]

Figure 3. SEM micrographs of zeolites synthesized by direct activation method.

The main disadvantage of the direct conversion method is relatively low-conversion degree. Several studies have focused on the relationship between composition of the raw materials and the synthesis conditions. However, conditions are not clearly established, and the combination of crystalline phases obtained is often unpredictable.

In **Figure 3**, the particles' morphology of zeolites synthesized by direct activation method is shown.

From **Figure 3**, it can be seen that ash morphology changes significantly, increasing mainly newly formed crystals on the surface of the particles. Following zeolitization, surface area increases significantly.

3.2. Fusion method followed by direct activation method

To obtain ash zeolites, the fusion step was proposed before the hydrothermal treatment. The ash mixed with solid NaOH is heated at high temperature, after which it is subjected to direct conversion. The alkaline fusion step before direct activation significantly improves zeolitization [45, 69].

Ash changing using the fusion method followed by direct activation has been reported by several authors [49, 69].

This method is known as the indirect method, as it consists of two phases:

- the fusion phase (450–600°C temperature, contact time 1–3 hours)

• the direct activation phase (temperature 60–100°C for 4–96 hours contact)

After the conclusion of the fusion phase, the materials were allowed time to age (room temperature, contact time of approximately 20 hours).

Material	References
Na-X, Na-A	[70]
Na-X	[49]
Na-A	[21, 49]
Na-Y	[47]
Na-X, Na-A, Na-P	[46, 67]
Fau	[47]
Zeolite 13X	[69]
ZFA	[48]
X, A, P, HS	[41]

Table 5. New materials obtained by the fusion method followed by direct activation.

Previous studies have also shown that the ash-synthesized materials by this method have a higher crystallinity and cation exchange capacity compared to the materials synthesized by the conventional hydrothermal method. The fusion method followed by direct activation method allows for a higher amount of alkaline extract ash silicates and alumina. The formation of new materials depends heavily on the SiO_2/Al_2O_3 ratio of the base material.

Table 5 presents the main types of new materials, which are obtained.

Figure 4 provides a few examples of new materials obtained by the fusion method followed by the direct activation.

[41] [71]

Figure 4. SEM images of materials obtained by fusion.

Obtaining new Na-A and Na-X type materials is also possible by direct activation, but when obtained by the fusion followed by direct activation method, these have higher cation exchange capacity and larger pores. The disadvantage of this method consists in the fact that the method involves the separation of the solid residues after a high content of silicon and aluminum has been dissolved in the alkaline solution. Disposal of the residues increases the possibility of obtaining a desired type of zeolite with high purity and regular particles, while leaving a new solid waste, with a very low production yield [45].

3.3. Ultrasound method

Ultrasounds (USs) are a form of mechanical energy that propagates in all directions in the form of beams. The ultrasound method has been reported primarily by Schmachtl et al. [72]. The ultrasound method is an indirect method, which is based on the investigation of the degree of interaction of the wave transmission ultrasonic properties with the precursor species of new materials, as they pass through the synthesis mixture. The method has been used in the synthesis of a wide range of nano- and microparticles.

Figure 5. SEM images obtained using ultrasound equipment. 3 - NaP zeolite synthesized by: (a) the sonochemical method at room temperature, 3 hours, (b) the sonochemical method followed by hydrothermal treatment for 12 hours at 100°C (c) sonochemical method followed by hydrothermal treatment for 24 hours at 100°C [78].

Of the various applications of the ultrasound treatment obtaining new ash-based materials requires special attention because it offers several advantages, the most significant being a reduced reaction time and crystallization temperature [21, 50, 51, 67, 73–75].

Feng et al. [76] showed that this method increases the compressive strength and thermal stability of the material. By accelerating the dissolution of the Al-Si, it improves polycondensation and the formation of semicrystalline phases. Ultrasound accelerates the chemical reactions of the components, the method being used to obtain new materials from sources, which contain Al and Si [72].

Ultrasound influences the physicochemical properties of the crystals, which result in following the synthesis. The effects of ultrasound are due to cavitation, a phenomenon may be defined as the occurrence of bubbles, which rise to the surface and then break. This can be explained by the fact that rapid successive expansions and compressions in liquid give rise to excessive stress in some areas.

The molar ratio of reactants, contact time, and temperature is important parameters that determine the type and properties of the new material. The disadvantage of this method is that the application of ultrasound changes the morphology of the crystals (crystal size decreases with power increase – **Figure 5**).

By applying different temperatures, concentrations of alkaline solution, the solid/liquid ratio, and the contact time, one can obtain materials with specific applications (**Table 6**).

Materials	References
4A	[51, 77, 79]
NaP	[78, 80]
MCM-22	[21]
X	[56]
Na-X	[81]

Table 6. New materials obtained by the ultrasound method.

3.4. Microwave method followed by hydrothermal activation

Querol et al. [36] obtained a zeolite using ash with a Si/Al ratio of 1.85 with a 62% amorphous phase, while the crystalline phases included: mullite (13%), magnetite (13%), quartz (7%), anhydrite (0.7%), and anorthite (0.3%). Two alkaline solutions of NaOH and KOH were used to obtain the zeolite crystals. The concentration of the solutions ranged from 1.0 to 5.0 M, and the temperature ranged between 150°C and 225°C, using both the conventional method and the microwave method (microwave oven with the available power of 1000 W). The crystallization time was reduced from conventional conversion of 8–98 hours to 10–30 minutes of microwave-assisted conversion. By changing the above-mentioned parameters, different stages of zeolite were obtained: Na-P1, hydroxy-sodalite, analcime, Linde F, tobermorite and kalsilit. Generally speaking, microwaves play a positive role in obtaining Na-P1. Experimentally, it was found that microwaves applied in the first 15 minutes, followed by 2 hours of synthesis led to the best results in as far as the ion-exchange capacity was concerned [53].

Materials	References
ZSM-11	[84]
Sodalite	[68, 85]
FAU	[86]
NaA	[83, 87]
4A, Na-P1	[88]

Table 7. New materials obtained by the ultrasound method.

Figure 6. SEM images of the materials obtained by the microwave method at different molar ratios of SiO_2/Al_2O_3: (a) 0.4, (b) 0.8, (c) 4.0, and (d) 10.0 [83].

Fukui et al. [82] used ash with a Si/Al ratio of 2.39 and NaOH at 100°C using the conventional method and the microwave method and obtained preponderantly filipsite. The particles of filipsite were smaller in the case of microwave treatment. Experimentally it has been demonstrated that microwaves increase the rate of nucleation, but result in a slower growth of the germ. The microwaves used for the first 2 hours, followed by conventional heating, positively influence the mechanism of germ formation and growth.

Tanaka et al. [83] used a 600 W microwave oven and the two-step synthesis method in order to obtain Na-A zeolite. The ashes used in the process had a Si/Al ratio of 2.76, a vitreous phase content of 49.4% and the NaOH solution had a concentration of 2.2 M. The mixture was irradiated in the oven for 60 minutes. The material obtained had an ion-exchange capacity of 4.70 meq/g, lower than the commercial material.

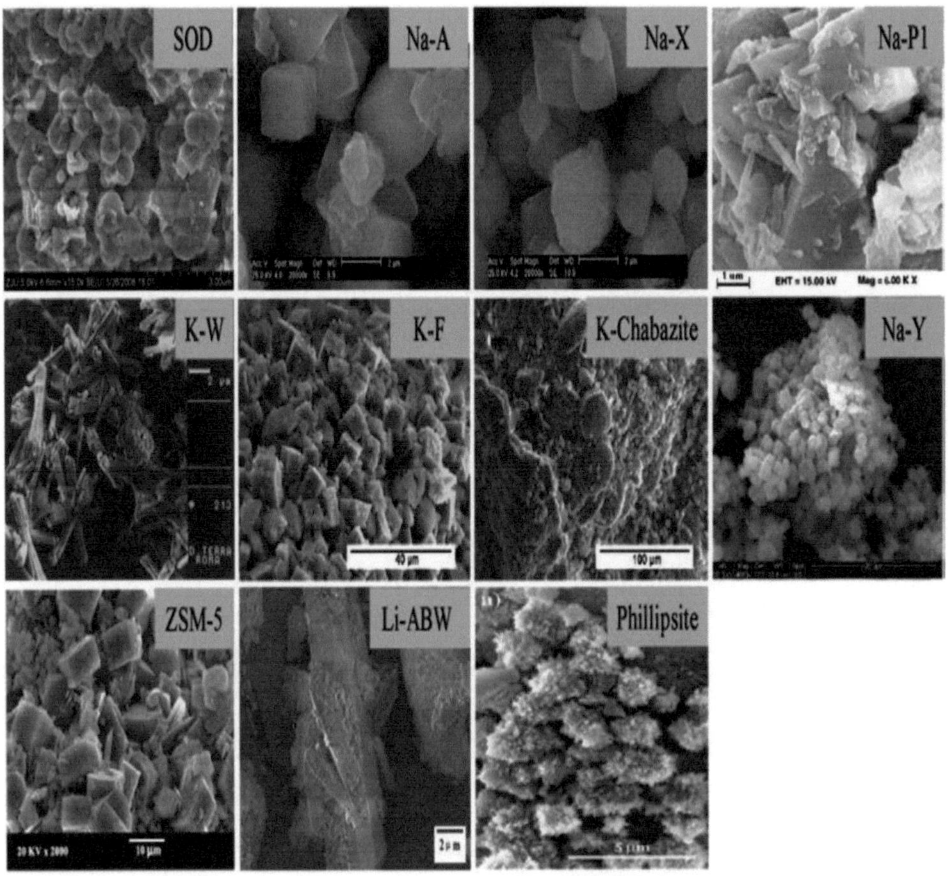

Figure 7. SEM images of zeolites synthesized from ashes [62, 67, 71, 74, 90, 91].

Kim and Lee [79] obtained Na-4A zeolite with an ion-exchange capacity of 5.5 meq/g. The literature indicates that microwave irradiation for 2 hours dissolves the compounds containing Si and Al and P zeolite and hydroxyl sodalite were obtained (**Table 7**). SEM images of the zeolites synthesized by the microwave method are shown in **Figure 6**.

Figure 7 presents the studies in the literature in the field of the materials obtained by modifying the ashes, depending on the morphology of the preponderant zeolite [62, 67, 71, 89, 90, 92]. **Figure 7** shows also that the morphology differs from one material to another from well-defined tetrahedral crystals to clusters of crystals.

Analysis of the literature data shows that alkaline attack of power plant ash allows one to obtain new materials, a mixture of zeolites, but the type and degree of zeolitization are influenced by the working conditions.

4. Conclusions

Zeolites from ashes are recognized as "intelligent fertilizer" because they have the ability to exchange cations allowing nutrients such as nitrogen, calcium, iron, magnesium, potassium, and water to be available in the plant and be eliminated slowly. The plants can absorb nutrients from the soil when required. Zeolites increase water reserves in soil needed during dry periods and are able to reversibly lose and retain water and change cations' network without changing the structure. They are chemically inert, nontoxic to plants, animals, humans, and the environment and regulate the balance of soil by immobilization of toxic contaminants to the plant.

The waste – fly ash can be capitalized by conversion in zeolites. The zeolites prepared by alkali activation of fly ash are aluminum-silicates and include three classes of inorganic polymers, depending on the ratio of silica/alumina. The composition of the mixture, and the reaction conditions such as SiO_2/Al_2O_3 ratio, hydroxide concentration, activation temperature, curing time, ratio of the solid/liquid, and pH significantly affect the formation and properties of zeolites. Limits of variation in these parameters are broad enough: ratio SiO_2/Al_2O_3 2–6, activation temperature 80–150°C, concentration NaOH or KOH solution 1–5 M, curing time of 4–48 hours and the ratio of the solid/liquid 1/1–1/4.

The methods used for the synthesis of new materials using solid waste – fly ashes are direct activation method (hydrothermal method), fusion method followed by direct activation method, ultrasound method, and microwave followed by direct activation method.

The main limitation of the synthesis process of zeolites is the relatively high temperature to dissolve Si and Al from the ash particles. If the temperature is reduced, then the yield of the synthesis is considerably reduced, and it is necessary to have a greater reaction time. However, KM, herschelite, K-chabazite, and Linde F can be obtained in the range of 125–200°C.

Author details

Maria Harja[1], Sorin Mihai Cimpeanu[2], Marcel Dirja[3] and Daniel Bucur[4*]

*Address all correspondence to: dbucur@uaiasi.ro

1 "Gheorghe Asachi" Technical University of Iasi, Faculty of Chemical Engineering and Environmental Protection, Iasi, Romania

2 University of Agronomical Sciences and Veterinary Medicine, Bucharest, Romania

3 University of Agricultural Sciences and Veterinary Medicine, Cluj Napoca, Romania

4 University of Agricultural Sciences and Veterinary Medicine, Iasi, Romania

References

[1] Ailincai C., Jitareanu G., Ailincai D., Zbant M. Evolution of main soil physical and chemical characteristics as influenced by tillage systems and fertilizers. Lucrari Stiintifice. Seria Agronomie, 2014;47:121–128.

[2] Savic R., Ondrasek G., Josimov-Dundjersk J. Heavy metals in agricultural landscapes as hazards to human and ecosystem health – a case study on zinc and cadmium in drainage channel sediments. Journal of the Science of Food and Agriculture, 2014;95(3): 466–470.

[3] Pearson M.S., Maenpaa K., Pierzynski G.M., Lydy M.J. Effects of soil amendments on the bioavailability of lead, zinc, and cadmium to earthworms. Journal of Environmental Quality, 2000;29:1611–1617.

[4] Bucur R.D., Cimpeanu C., Barbuta M., Ciobanu G., Paraschiv G., Harja M. A comprehensive characterization of ash from Romania thermal power plant. Journal of Food, Agriculture & Environment, 2014;12(2):943–949.

[5] Ahmaruzzaman M. A review on the utilization of fly ash. Progress Energy Combustion Science, 2010:36(3):327–363.

[6] Ciocinta R.C., Harja M., Bucur D., Rusu L., Barbuta M., Munteanu C. Improving soil quality by adding modified ash. Environmental and Engineering Management Journal, 2012;11(2):297–305.

[7] Majid A., Argue S. Remediation of heavy metal contaminated solid wastes using agglomeration techniques. Minerals Engineering, 2001;14(11):1513–1525.

[8] Lin C.F., Lo S.S., Lin H.Y., Lee Y.C. Stabilization of cadmium contaminated soils using synthesized zeolite. Journal of Hazardous Materials, 1998:60:217–226.

[9] Adriano D.C. Trace Elements in Terrestrial Environments; Biogeochemistry, Bioavailability and Risks of Metals, 2nd edn, Springer, New York, 2001; 866 pp.

[10] Ciocinta R.C., Cîmpeanu S.M., Bucur R.D., Harja M., Barbuta M., Gurita A.A. Potential applications in agriculture of new materials synthesized from ash. Scientific Papers. Series E. Land Reclamation, 2012;101–104.

[11] Chmielewská E. Abundantly available natural or bioinspired materials for aqueous pollutants removal. Current Green Chemistry, 2015;2(4):371–378.

[12] Diels L., van der Lelie N., Bastiaens L. New developments in treatment of heavy metal contaminated soils. Reviews in Environmental Science Bio/Technology, 2002;1:75–82.

[13] Ciocinta R.C., Harja M., Bucur D., Buema G. Optimization of the conditions for conversion of ash into zeolite materials. Journal of Food, Agriculture & Environment, 2013;11(1):1108–1012.

[14] Fansuri H., Pritchard D., Zhang D. Manufacture of low grade zeolites from fly ash for fertilizer application. 2008; Research Report 91.

[15] Ma L.Q., Tcaina S.J., Logan T.J. In situ lead immobilization by apatite. Environmental Science and Technology, 1993;27:214–220.

[16] Cao X.R., Ma L.Q., Chen M., Singh S.P., Harris W.G. Phosphate-induced metal immobilization in a contaminated site. Environmental Pollution, 2003;122:19–28.

[17] Ciccu R., Ghiani M., Serci A., Fadda S., Peretti R., Zucca A. Heavy metal immobilization in the mining-contaminated soils using various industrial wastes. Minerals Engineering, 2003;16:187–192.

[18] Guo G., Zhou Q., Ma L. Availability and assessment of fixing additives for the in situ remediation of heavy metal contaminated soils: a review. Environmental Monitoring and Assessment, 2006;116:513–528.

[19] Bolan N.S., Adriano D.C., Mani A.P., Duraisamy A. Immobilization and phytoavailability of cadmium in variable charge soils. II. Effect of lime compost. Plant and Soil, 2003;251:187–198.

[20] Chen M., Ma L., Singh S., Cao X., Melamed R. Field demonstration of in situ immobilization of soil Pb using P amendments. Advanced Environmental Research, 2003;8:93–102.

[21] Wang S., Terdkiatburana T., Tade M.O. Single and co-adsorption of heavy metals and humic acid on fly ash. Separation and Purification Technology, 2008;58:353–358.

[22] Ciccu R., Ghiani M., Peretti R., Serci A., Zucca A. Heavy metal immobilization using fly ash in soils contaminated by mine activity. In: International Ash Utilization Symposium, May, 2001. Center for Applied Energy Research, University of Kentucky, Paper #6.

[23] Buema G., Cimpeanu S.M., Sutiman D.M., Bucur R.D., Rusu L., Cretescu I., Ciocinta R.C., Harja M. Lead removal from aqueous solution by bottom ash. Journal of Food, Agriculture & Environment, 2013;11(1):1137–1141.

[24] Deja J. Immobilization of Cr^{6+}, Cd^{2+}, Zn^{2+} and Pb^{2+} in alkali-activated slag binders. Cement Concrete Research, 2002;32:1971–1979.

[25] Albino V., Cioffi R., Marroccoli M., Santoro L. Potential application of ettringite generating systems for hazardous waste stabilization. Journal of Hazardous Materials, 1996;51:241–252.

[26] Industrial Minerals.net. Zeolite Soil Conditioner. 2006. July 18.

[27] Virta R.L. Zeolites. 2002. U.S. Geological Survey Minerals Yearbook.

[28] Mumpton F.A. La roca magica: uses of natural zeolites in agriculture and industry. Proceedings of the National Academy of Sciences of the United States of America, 1999;96:3463–3470.

[29] Rota Mining Co. Natural Zeolite as Soil Conditioner. 2006. July 18, 2006.

[30] Zeolite.ca.NorthAmerica'sZeolite. 2006; http://www.zeolite.ca/SoilConditioner-AG.htm.

[31] Treacy M.M.J., Higgins J.B. Collection of Simulated XRD Powder Patterns for Zeolites, 2001.

[32] Harja M., Suteu D., Ciobanu G., Rusu L. Materials based on ash for environmental protection. I. Obtaining and characterization, Scientific Researches, Agronomy Series, 2007;50:14–22.

[33] Criado M., Fernandez-Jimenez A., Palomo A. Alkali activation of fly ash: effect of the SiO_2/Na_2O ratio Part I: FTIR study. Microporous and Mesoporous Materials, 2007;106:180–191.

[34] Izquierdo M., Querol X. Leaching behaviour of elements from coal combustion fly ash: an overview. International Journal of Coal Geology, 2001;94:54–66.

[35] Höller H., Wirsching U. Zeolites formation from fly ash. Fortschritte der Mineralogie, 1985;63:21–43.

[36] Querol X., Alastuey A., Lopez-Soler A., Plana F., Andres J.M., Juan R., et al. A fast method for recycling fly ash: microwave-assisted zeolite synthesis. Environmental Science and Technology, 1997;31:2527–2533.

[37] Inada M., Eguchi Y., Enomoto N., Hojo J. Synthesis of zeolites from coal fly ash with different silica-alumina composition. Fuel, 2005;84:299–304.

[38] Juan R., Hernandez S., Andres J.M., Ruiz C. Synthesis of granular zeolitic materials with high cation exchange capacity from agglomerated coal fly ash. Fuel, 2007;86:1811–1821.

[39] Izidoro J.C., Fungaro D., Santos F.S., Wang S. Characteristics of Brazilian coal fly ashes and their synthesized zeolites. Fuel Process Technology, 2012;97:38–44.

[40] Liu H., Peng S., Shu L., Chen T., Bao T., Frost R. Magnetic zeolite NaA: synthesis, characterization based on metakaolin and its application for the removal of Cu^{2+}, Pb^{2+}. Chemosphere, 2013;91:1539–1546.

[41] Zhang X., Tang D., Jiang G. Synthesis of zeolite NaA at room temperature: the effect of synthesis parameters on crystal size and its size distribution. Advanced Powder Technology, 2013;24(3):689–696.

[42] Sharma P., Sharma M., Tomar R. Na-HEU zeolite synthesis for the removal of Th(IV) and Eu(III) from aqueous waste by batch process. Journal of the Taiwan Institute of Chemical Engineers, 2013;44:480–488.

[43] Shoumkova A., Stoyanova V. Zeolites formation by hydrothermal alkali activation of coal fly ash from thermal power station "Maritsa 3", Bulgaria. Fuel, 2013;103:533–541.

[44] Visa M. Synthesis and characterization of new zeolite materials obtained from fly ash for heavy metals removal in advanced wastewater treatment. Powder Technology, 2016:294(6):338–347.

[45] Hollman G.G., Steenbruggen G., Janssen-Jurkovicova M. A two-step process for the synthesis of zeolites from coal fly ash. Fuel, 1996;78:1225–1230.

[46] Rayalu S., Meshram S.U., Hasan M.Z. Highly crystalline Faujasitic zeolites from fly ash. Journal of Hazardous Materials, 2000:B77:123–131.

[47] Rios C., Williams C., Roberts C. A comparative study of two methods for the synthesis of fly ash-based sodium and potassium type zeolites. Fuel, 2009;88:1403–1416.

[48] Javadian H., Ghorbani F., Tayebi H., Asl S.M.H. Study of the adsorption of Cd (II) from aqueous solution using zeolite-based geopolymer, synthesized from coal fly ash; kinetic, isotherm and thermodynamic studies. Arabian Journal of Chemistry, 2015;8:837–849.

[49] Izidoro J., Fungaro D., Abbott J., Wang S. Synthesis of zeolites X and A from fly ashes for cadmium and zinc removal from aqueous solutions in single and binary ion systems. Fuel, 2013;103:827–834.

[50] Park J., Kim B.C., Park A.S., Park H.C. Conventional versus ultrasonic synthesis of zeolite 4A from kaolinite. Journal of Materials Science Letters, 2001;20:531–533.

[51] Andac O., Tather M., Sirkecioglu A., Ece I., Erdem-Senatalar A. Effects of ultrasound on zeolite A synthesis. Microporous Mesoporous Materials, 2005;79:225–233.

[52] Murayama N., Yamamoto H., Shibata J. Mechanism of zeolite synthesis from coal fly ash by alkali hydrothermal reaction. International Journal of Mineral Processing, 2002;64(1):1–17.

[53] Bukhari S.S., Behin J., Kazemian H., Rohani S. Conversion of coal fly ash to zeolite utilizing microwave and ultrasound energies: a review. Fuel, 2015;140:250–266.

[54] Molina A, Poole C. A comparative study using two methods to produce zeolites from fly ash. Mineral Engineering, 2004;17:167–173.

[55] Moreno N., Querol X., Plana F., Andres J.M., Janssen M., Nugteren H. Pure zeolite synthesis from silica extracted from coal fly ashes. Journal of Chemical Technology and Biotechnology, 2002;77:274–279.

[56] Belviso C., Giannossa L., Javier Huertas F., Lettino A., Mangone A., Fiore S. Synthesis of zeolites at low temperatures in fly ash-kaolinite mixtures. Microporous and Mesoporous Materials, 2015;212(8):35–47.

[57] Querol X., Umana J.C., Plana F., Alastuey A., Lopez-Soler A., Medinaceli A., et al. Synthesis of zeolites from fly ash at pilot plant scale. Examples of potential applications. Fuel, 2001;80:857–865.

[58] Moriyama R., Takeda S., Onozaki M., Katayama Y., Shiota K., Fukuda T., et al. Large-scale synthesis of artificial zeolite from coal fly ash with a small charge of alkaline solution. Fuel, 2005;84:1455–1461.

[59] Moutsatsou A., Karakasia O.K., Koukouzas N., Itskos G.S., Vasilatosc C. Synthesis of zeolitic materials utilizing CFB-derived coal fly ash as a raw material. WanteEng-2nd International Conference on Engineering for Waste Valorisation, 2006. Patras.

[60] Harja M., Buema G., Bulgariu L., Bulgariu D., Sutiman D.M., Ciobanu G. Comparative study for removal of cadmium (II) onto low cost adsorbents. Korean Journal of Chemical Engineering, 2015;32(9):1804–1811.

[61] Wu D., Zhang B., Yan L., Kong H., Wang X. Effect of some additives on synthesis of zeolite from coal fly ash. International Journal of Mineral Processing, 2006;80:266–272.

[62] Chareonpanich M., Namto T., Kongkachuichay P., Limtrakul J. Synthesis of ZSM-5 zeolite from lignite fly ash and rice husk ash. Fuel Processing Technology, 2004;85:1623–1634.

[63] Querol X., Moreno N., Umaña J.C., Alastuey A., Hernandez E., Lopez-Soler A., Plana F. Synthesis of zeolites from coal fly ash: an overview. International Journal of Coal Geology, 2002;50:413–423.

[64] Chareonpanich M., Jullaphan O., Tang C. Bench-scale synthesis of zeolite A from subbituminous coal ashes with high crystalline silica content. Journal of Cleaner Production, 2011;19(1):58–63.

[65] Ali I.O., Hassan A.M., Shaaban S.M., Soliman K.S. Synthesis and characterization of ZSM-5 zeolite from rice husk ash and their adsorption of Pb^{2+} onto unmodified and surfactant-modified zeolite. Separation and Purification Technology, 2011;83:38–44.

[66] Izidoro J.C., Fungaro D., Wang S. Zeolite synthesis from Brazilian coal fly ash for removal of Zn^{2+} and Cd^{2+} from water. Advanced Materials Research, 2012;356–360:1900–1908.

[67] Harja M., Buema G., Sutiman D.M., Cretescu I. Removal of heavy metal ions from aqueous solutions using low-cost sorbents obtained from ash. Chemical Papers, 2013;67(5):497–508.

[68] Noli F., Buema G., Misaelides P., Harja M. New materials synthesized from ash under moderate conditions for removal of toxic and radioactive metals. Journal of Radioanalytical and Nuclear Chemistry, 2015;303(3):2303–2311.

[69] Mishra T., Tiwari S.K. Studies on sorption properties of zeolite derived from Indian fly ash. Journal of Hazardous Materials, 2006;B137:299–303.

[70] Purnomo C.W., Salim C., Hinode H. Synthesis of pure Na-X and Na-A zeolite from bagasse fly ash. Microporous and Mesoporous Materials, 2012;162:6–13.

[71] Kazemian H., Naghdali Z., Ghaffari Kashani T., Farhadi F. Conversion of high silicon fly ash to Na-P1 zeolite: alkaline fusion followed by hydrothermal crystallization. Advanced Powder Technology, 2010;21(3):279–283.

[72] Schmachtl M., Kim T.J., Grill W., Herrmann R., Scharf O., Schwieger W., Schertlen R., Stenzel C. Ultrasonic monitoring of zeolite synthesis in real time. Ultrasonics, 2000;38:809–812.

[73] Curteanu S., Buema G., Piuleac C.G., Sutiman D.M., Harja M. Neuro-evolutionary optimization methodology applied to the synthesis process of ash based adsorbents. Journal of Industrial and Engineering Chemistry, 2014;20(2):597–604.

[74] Harja M., Bucur D., Cimpeanu S.M., Ciocîntă R.C. Conversion of ash on zeolites for soil application. Journal of Food, Agriculture & Environment, 2012;10(2):1056–1059.

[75] Harja M., Buema G., Sutiman D.M., Munteanu C., Bucur D. Low cost adsorbents obtained from ash for copper removal. Korean Journal of Chemical Engineering, 2012;29(12):1735–1744.

[76] Feng D., Tan H., Van Deventer J.S.J. Ultrasound enhanced geopolymerisation. Journal of Materials Science, 2004;39:571–580.

[77] Musyoka N., Petrik L., Hums E., Baser H., Schwieger W. In situ ultrasonic monitoring of zeolite A crystallization from coal fly ash. Catalysis Today, 2012;190(1):38–46.

[78] Pal P., Das J.K., Das N., Bandyopadhyay S. Synthesis of NaP zeolite at room temperature and short crystallization time by sonochemical method. Ultrasonics Sonochemistry, 2013;20(1):314–321.

[79] Kim J.K., Lee H.D. Effects of step change of heating source on synthesis of zeolite 4A from coal fly ash. Journal of Industrial and Engineering Chemistry, 2009;15(5):736–742.

[80] Noli F., Buema G., Misaelides P., Harja M. Retention of radiocesium from aqueous solutions using modified power plant ash. Journal of Radioanalytical and Nuclear Chemistry, 2015, 1–8, doi 10.1007/s10967–015–4611–6

[81] Erten-Kaya Y., Cakicioglu-Ozkan F. Effect of ultrasound on the kinetics of cation exchange in NaX zeolite. Ultrasonics Sonochemistry, 2012;19(3):701–706.

[82] Fukui K, Arai K, Kanayama K, Yoshida H. Phillipsite synthesis from fly ash prepared by hydrothermal treatment with microwave heating. Advanced Powder Technology, 2006;17:369–832.

[83] Tanaka H., Fujii A., Fujimoto S., Tanaka Y. Microwave-assisted two-step process for the synthesis of a single-phase Na-A zeolite from coal fly ash. Advanced Powder Technology, 2008;19(1):83–94.

[84] Jury F.A., Polaert I., Estel L., Pierella L.B. Enhancement of synthesis of ZSM-11 zeolite by microwave irradiation. Microporous and Mesoporous Materials, 2014;198(1):22–28.

[85] Xu X., Bao Y., Song C., Yang W., Liu J., Lin L., Microwave-assisted hydrothermal synthesis of hydroxy-sodalite zeolite membrane. Microporous and Mesoporous Materials, 2004;75(3):173–181.

[86] Cheng Z., Han S., Sun W., Qin Q. Microwave-assisted synthesis of nanosized FAU-type zeolite in water-in-oil microemulsion. Materials Letters, 2013;95(3):193–196.

[87] Ansari M., Aroujalian A., Raisi A., Dabir B., Fathizadeh M. Preparation and characterization of nano-NaX zeolite by microwave assisted hydrothermal method. Advanced Powder Technology, 2014;25(2):722–727.

[88] Inada M., Tsujimoto H., Eguchi Y., Enomoto N., Hojo J. Microwave-assisted zeolite synthesis from coal fly ash in hydrothermal process. Fuel, 2005;84(12–13):1482–1486.

[89] Harja M., Rusu L., Bucur D., Ciocinta R.C. Fly ash-derived zeolites as adsorbents for Ni removal from waste water. Revista Romana de Chimie, 2012;57(6):587–597.

[90] Belviso C., Cavalcante F., Lettino A., Fiore S. Effects of ultrasonic treatment on zeolite synthesized from coal fly ash. Ultrasonics Sonochemistry, 2011;18(2):661–668.

[91] Yao Z.T., Ji X.S., Sarker P.K., Tang J.H., Ge L.Q., Xia M.S., Xi Y.Q. A comprehensive review on the applications of coal fly ash. Earth-Science Reviews, 2015;141:105–121.

[92] Harja M., Advanced capitalization of the fly ash to obtain new materials, 2016, Thesis of Habilitation, 205

Chapter 4

Fly Ash Derived Zeolites in the Removal of Toxic Compounds

Magdalena Wdowin and Wojciech Franus

Additional information is available at the end of the chapter

http://dx.doi.org/10.5772/64362

Abstract

The present paper describes the previous obtained results concerning purification of gases, water and sewages from toxic components such as mercury, radionuclides and petroleum substances. The tested materials constitute synthetic zeolites obtained in hydrothermal reaction, i.e., Na-X, Na-P1, and natural zeolite clinoptilolite. The zeolites (Na-X, Na-P1 and clinoptilolite) needed for mercury capture were activated by silver ions; however, for the removal of petroleum substances (Na-X and Na-P1) as well as radionuclides (Na-P1) raw zeolite materials were used. The review of the results have shown that zeolites used both for mercury capture and the removal of radionuclides, as well as petroleum substances, can be considered promising sorbent materials.

Keywords: fly ash, na-X, na-P1, clinoptilolite, mercury, radionuclides, petroleum substances

1. Introduction

The chemical impact on the environment as well as human health constitutes an increasing concern [1]. Therefore, the awareness of relationships between ecosystem and human health and their distribution in the environment [2] is growing worldwide. Considering chemicals that are toxic and harmful for the organisms, Plant et al. [1] distinguished four main groups of pollutants, i.e., (1) inorganic elements such as heavy metals, i.e., As, Cd, Hg, or Pb but also Se, I, NO_x, Ga and PGEs; (2) radioactive substances such as ^{238}U, ^{226}Ra, ^{222}Rn as well as process materials, i.e., depleted uranium or isotopes ^{137}Cs, ^{90}Sr, ^{243}Am and ^{240}Pu; (3) persistent organic pollutants such as dichlorodiphenyltrichloroethane (DDT), polychlorinated biphenyls (PCBs), polybrominated diphenyl ethers (PBDEs), perfluorooctane sulphonate (PFOS); and (4) human and veterinary pharmaceuticals.

The European Environment Agency and the United Nations Environment Programme articulated the problem of chronic exposure to potentially toxic chemicals [3]. The increasing concern about harmful compounds in the environment is also reflected by the growing number of European Union (EU) Directives and national and international conventions related to chemicals [1].

For the removal of toxic pollutants in the environment, many sorbents and materials are investigated. One of the promising materials considered over the years is zeolites (alumino-silicates of alkali and earth metals). Due to their crystalline, infinite and three-dimensional crystal structure, they possess many important properties that are helpful in environmental and chemical sciences among others, such as, adsorption, dehydratation-rehydratation, ion-exchange molecular sieve and catalytic properties [4].

The various potential applications of both synthetic and natural zeolites depend on their fundamental physical and chemical properties that are in turn directly related to the chemical composition and crystal structure of individual species [5].

The present work describes the evaluation of zeolite materials for three toxic compounds, namely, mercury from flue gases, and radionuclides from mine waters and petroleum substances.

2. Zeolites as mercury sorbents

2.1. Characteristic of mercury pollutions in flue gas

Mercury in various forms (for example, Hg^0, HgCl and HgO) is one of the most important and dangerous environmental contaminants and has received special attention because of its high volatility, high toxicity and bioaccumulation ability [6].

It is believed that coal combustion is one the main sources of mercury emissions in the atmosphere, accounting for 60%, or even more, of the total mercury emissions [7]. The rest of the emissions originate from cement clinker production, and primary production of ferrous and non-ferrous metals is a predominant source of global anthropogenic mercury emission. Annual emissions of mercury to air form current anthropogenic sources are responsible for about 30% [8].

Considering combustion of fuels, the mercury in coal is volatilized and converted into elemental mercury (Hg^0) vapour in the high-temperature regions of coal-fired boilers. During cooling of flue gas, a series of complex reactions take place converting Hg^0 into ionic mercury (Hg^{2+}) compounds and/or Hg compounds (Hg_p) that are in a solid phase at flue gas cleaning temperatures or Hg that is adsorbed onto the surface of other particles. The presence of chlorine in flue gas causes the formation of mercuric chloride ($HgCl_2$) at flue gas cleaning temperatures. Nevertheless, Hg^0 is speciation; as a result, Hg enters the flue gas cleaning device(s) as a mixture of Hg^0, Hg^{2+} and Hg_p. It is thought that majority of gaseous mercury in bituminous coal-fired boilers constitutes Hg^{2+}. But, the majority of gaseous mercury in subbituminous- and lignite-fired boilers is Hg^0 [9].

In conclusion, mercury in flue gas is found in three major chemical forms: gaseous elemental mercury (Hg^0), gaseous oxidized mercury (Hg^{2+}) and particulate-bound mercury (Hg_p). The mercury amount adsorbed on the particles of ash ranges from 5 to 10% of the total quantity of Hg released during the combustion processes. The remaining 90–95% of the mercury is released in the gaseous form. Mercury concentrations of flue gas from coal combustion generally range from 5 to 10 μg/m^3, with a large range in the relative proportions of Hg^0, Hg^{2+} and Hg_p. The most dangerous and the most stable mercury form that accounts for over 90% of the total mercury in the atmosphere is the elemental form. It is hazardous due to long residence time in the atmosphere (from 6 months to 2 years) but it could also be as short as hours to weeks under specific environmental conditions [10, 11] and due to migrating hundreds of kilometres from the source of emission. The gaseous form of Hg^{2+} lasts for only several days to several weeks in the atmosphere [12–14].

Another problem is that mercury is insoluble in the water, thus making inefficient the existing methods of their removal from flue gases [15]. To remove mercury-activated carbons (AC) commercially, their high cost and instability of the structure (degradation) in the exhaust gas temperatures are the major problems. In addition, these materials are not very efficient for removing mercury in the elemental form. Consequently, a large number of studies have focused on the possibility of using new techniques based on zeolites for removing gaseous elemental mercury species [16, 17]. The other forms of mercury (oxidized and gaseous divalent mercury Hg^{2+}) can react with ash particles to form the solids. These two forms, as compared to Hg^0, are efficiently removed from the gas stream by the conventional purification methods (wet desulphurization or bag filter) [14, 18–21].

2.2. Sorption mechanism of Hg on zeolites

Many materials are under examination because of their potential oxidation, capture and catalysis of mercury. The most frequently studied are, among others, palladium, gold, iridium, platinum, iron, fly ash and activated carbons [22].

Previous studies on removing gaseous forms of mercury (mainly Hg^0) from zeolites showed negligible adsorption of mercury vapour compared with the commercially used activated carbons. This may be related to the lack of chemical affinity of zeolites to mercury or a stronger affinity to the other, main components of exhaust gas (including water vapour present in the exhaust gas, which is rapidly adsorbed in the zeolite structure). Therefore, it can be concluded that the adsorption of gaseous forms of mercury is not a pure physical adsorption. In addition to zeolites, sulphur-activated carbons did not effectively adsorbed mercury, whereas sulphur-impregnated activated carbons are good sorbents of Hg and currently they are commercially used [23].

Silver activation of zeolite materials significantly improves their adsorption capacity with respect to mercury [17, 24–26], which is related to chemical affinity of silver to mercury. The magnetic properties of Fe-activated zeolites also significantly influence the adsorption of mercury [27]. Manganese-oxidized mercury, Hg^0 to HgO, compounds are more easily adsorbed [28]. On the other hand, silver has a high affinity for mercury [17]. Mercury in contact with the silver-activated zeolite is bonded to the zeolite structure as a silver amalgam. Then, mercury is released from the structure of zeolite during the heating of zeolite to a temperature of about 400°C, indicating the possibility of regeneration of the material. An example for silver-impregnated zeolites is that the magnetic composite zeolite-containing silver nanoparticles were able to capture mercury from the flue gases in large scale from coal-fired power plants [26]. In order to improve the sorption of mercury on different sorbents, the bromine chloride (BrCl) to oxidize Hg^0 to $HgCl_2$ was used [29]. Nowadays, the introduction of selective catalytic reduction (SCR) increases the efficiency of the oxidation of Hg^0 to Hg^{2+} [30].

2.3. Results of mercury capture

Considering the results obtained by other authors, so far many materials have been observed and studied for their potential to remove mercury from exhaust gases. Others activated carbons (impregnated with iodine, bromine and sulphur) and carbon-based materials are fly ash, metals (such as silver, gold, copper and lead), and their oxides and sulphides, calcium species (lime) and zeolites [10, 15, 21, 31–35]. Unfortunately, most of these sorbents are less effective at higher temperature, in addition they have low capacities, cannot be regenerated and are easily deactivated by flue gas components such as sulphur oxides (SO_x) [36, 37], which means that the search for the ideal mercury sorbent is far from over [38].

Other raw materials for the capture of mercury zeolites were tested both synthetically and naturally. However, compared with the activated carbon, they showed much less adsorption capacity relative to the gaseous form of mercury. Therefore, significant attempts have been taken to activate them in order to improve their adsorption capacity.

The sorption tests of mercury compounds were performed by Wdowin et al. [17] at laboratory scale in pure mercury stream wherein nitrogen was used as carrier gas. Similar experiments were carried out by Wdowin et al. [6] in flue gas stream wherein gas mixtures were composed of SO_x, NO_x, CO_2, HCl, HF and CO.

The tested materials in the experiments carried out by Wdowin et al. [17] constitute Na-X, Na-P1, clinoptilolite. In order to enhance their sorption capacity to mercury, the investigated materials were activated with silver by using the ion-exchange method.

The first experiment was carried out in a pure stream of Hg^0 wherein nitrogen was used as carrier gas. The tested raw zeolite materials were not able to remove more than 10% (by weight) of mercury; therefore, their sorption capacity at the breakthrough of the bed could not be calculated.

Table 1 clearly shows that the addition of Ag^+ to the tested samples significantly improves the performance of the zeolite with respect to mercury capture. The best results were observed for the sample of Ag-X, which is associated with the highest BET surface area and the largest volume of micro- and mesopores. This result suggests that both micro- and mesopores are essential for the mercury capture. Moreover, in this case multilayer adsorption is an important parameter, but only to some extent, because Ag-P1 possesses a larger volume of meso- and macropores and Ag^+ ions, but this zeolite captures lower amount of mercury. The results were compared with commercially available activated carbons.

Sorbents	Flow rate N_2 (ml/min)	Evaporation (mg/min)	Time to breakthrough (5%)* (min)
Na-X	80	0.00288	2
Ag-X	80	0.00288	13017
Ag-X	160	0.005766	5327
Na-P1	80	0.00286	0
Ag-P1	80	0.002883	1536
Clinoptilolite	80	0.00286	2
Ag-Clinoptilolite	80	0.00286	5
AC/Br	80	0.002883	2538

Breakthrough when 5% of Hg passed through the sorbent bed without being captured.

Table 1. The results of Hg^0 capture on the raw and silver-activated zeolites [17].

From **Table 1** it can be concluded that Ag-X zeolite found very promising results because it has achieved five times better result of Hg capture as compared with AC/Br zeolite material.

Moutsatsou et al. [39] obtained similar results of Hg capture on clinoptilolite; it was very low especially compared with the synthetic zeolites. It also was not improved by silver-activated clinoptilolite.

The removal of the gaseous forms of mercury from exhaust gases was carried out by Wdowin et al. [6] on the zeolite Na-X-activated silver (Ag-X) in an especially designed oven. The tested zeolite beds were in the granular and powder form. The Hg flow rate was established at 85 dm^3/h. The estimated exhaust gas composition at an oxygen concentration of the order of 4–8% was as follows: CO_2 = 11–17%, CO = 20–50 ppm, NO = 70–100 ppm, SO_2 = 650–1000 ppm. The fixed gas stream was passed through a sorbent bed.

Studies that remove gaseous forms of mercury from the exhaust gas at the Ag-X zeolite in the form of powder and granulate showed that granulation enhances the efficiency of the tested material. The average penetration of mercury was 15% in the test lasting about two hours, whereas the presence of other components (mainly sulphur compounds in the flue gases) reduces the efficiency of the tested sorbent in comparison to tests carried out in a clean gas stream of Hg (carrier gas N_2). The better results were obtained for granulated zeolite materials.

Hsi et al. [40] studied fly ash and zeolites of types A and Y in terms of the removal of elemental Hg and $HgCl_2$ in the flue gases. They studied both starting materials and sulphur-activated materials. Studies have shown negligible adsorption of Hg compounds by initial fly ash and sulphur-activated materials. These results explained that the adsorbents' porous structure, mainly as micropores, is one of the important factors for Hg^0 and $HgCl_2$ adsorption. In addition, the authors point out that the reason for these results is the water content in the structure of the zeolite. Therefore, prior to impregnation it is necessary to get rid of H_2O from the zeolite structure. Studies show that too small size of the zeolite channels is also a cause of low adsorption of Hg compounds.

Morency et al. [41] investigated the use of a low-cost additive to improve the capability of a high surface area sorbent such as zeolite to remove all forms of mercury vapour from power plant flue gases. They examined treated zeolite wherein preliminary research led to the conclusion that a zeolite material could achieve substantial mercury removal at a significant cost reduction over current mercury control technologies. They carried out the adsorption test of Hg compounds at 150°C on activated natural zeolite (clinoptilolite) and synthetic (A, X, Y, ZSM-5) zeolites. The results proved to be similar to the results obtained on the active carbons. In addition, the zeolites used in the experiments captured all forms of mercury present in the flue gas. This material is also much cheaper than the commercially used activated carbons for this purpose. These authors also found that synthetic zeolites are better for the removal of mercury compounds due to the constant size of the channels and chambers, whereas sizes of natural zeolites are variable and are often associated with a number of network defects; however, natural zeolites are much cheaper material. In addition, after the Hg adsorption experiment on zeolites, these authors conducted leachability test on this material wherein they obtained positive results, i.e., the tested sorbent can be safely disposed in a landfill if sale of the ash is not an option.

In addition, Morency et al. [42] conducted a successful Hg adsorption test on two types of treated zeolites. The experiment was conducted at a temperature of 130–200°C of flue gas and a concentration of 30–60 $\mu m/m^3$ of Hg. Studies have shown almost 100% adsorption of Hg at an Hg ratio of 25,000.

Morency et al. [16] carried out experiment on adsorption of Hg compounds of the flue gases on activated carbons and raw and treated zeolites using appropriate chemical reagents. The results showed that in the case of treated zeolites, they show similar results to those of active carbons whereas their efficiency is greater than 90% at a ratio of injected mercury 25,000 g/g.

Pavlish et al. [43] have studied the adsorption of mercury compounds wherein A class of magnetic zeolite composites with supported silver nanoparticles has been tested for elemental mercury removal from power plant flue gas. As a result, this type of activation mercury capture levels of 45–92% was achieved at high ratios of sorbent to mercury ranging from 20,000 to 100,000, with no decrease in capture observed and an increase in gas temperature between 130 and 200°C. In Addition, high dose rates of zeolites, which are composed of aluminosilicate, will not degrade the commercial value of fly ashes used as a substitute for cement in concrete or as filler in plastics.

In many reports on the removal of mercury from flue gas, it is underlined that zeolite-based sorbents are not as efficient as activated carbons but can be much more economical to use than the commonly used carbons.

Mercury sorption experiments on the non-carbon materials at 110°C were conducted by Jurng et al. [23]. For this purpose, they used zeolite and bentonite materials with a relatively low surface area (less than 30 m^2/g), which were sulphur activated in order to improve the sorption. As an activating agent, CS_2 solution was used. The results showed 50% of the sorption of Hg compounds for both the starting and activated zeolites and bentonites.

3. Zeolites as radionuclide sorbents

3.1. Characteristics of radionuclides in mine water

The development in the field of nuclear science and technology, generation of electricity in nuclear power plants, nuclear weapons testing, and the use of radionuclides in a research and medicine have led to the increase in the production of various wastes (both liquid and solid) containing radionuclides and release of artificial radionuclides into the environment. It is thought that radiocaesium ^{134}Cs and ^{137}Cs are one of the most abundant radionuclides in nuclear fission products and is considered as hazardous element [44, 45].

One of the most important sources of radionuclide pollution is the mine waters. Some mines have a serious problem with utilization of mine water rich in the radionuclide elements among others Silesia mines in Poland. There are also other areas all over the world where significant radioanomalies are present, including a coal mining district in Germany where radium-bearing waters occurred with ^{226}Ra concentration up to 63 $kBq \cdot m^{-3}$. One of the major sources of radiation in such areas is related with hard coal mining activity. Sometimes in radium-bearing mine water barium ions are also present, in some mines in concentrations of up to 2 $g \cdot dm^{-3}$. Contrary, in other waters no barium can be found but radium and sulphate ions are present. The presence of barium in mine waters is the most important factor for the further behaviour of radium isotopes in mine galleries or on the surface. From waters with radium and barium these elements always co-precipitate as sulphates, when such waters are mixed with any water, containing sulphate ions [46]. Coal, deep circulation groundwater, and deposits contain significant quantities of ^{226}Ra and ^{228}Ra as a member of the uranium chain.

In addition, natural waters with similarly high radium concentrations are observed, especially in oil fields. For example, hot springs in Iran contain up to 330 $kBq \cdot m^{-3}$ of ^{226}Ra [47].

3.2. Sorption mechanisms of radionuclides on zeolites

In the case of radionuclide sorption onto zeolite such mechanisms are quite simple in comparison to mercury capture. Zeolites due to their ion-exchange properties have received great attention, especially for application in radioactive liquid waste treatment [48, 49]. The radioisotopes in the form of cations, present in the liquid effluents of low and intermediate levels of liquid waste, are removed by the ion exchange with the Na^+ ions present in the zeolite structure in ion-exchange positions. Zeolite materials are characterized by high exchange capacity, selectivity and specificity, and good resistant to radiation, which is very important with respect to the immobilization of radionuclides and final disposal when compared with organic ion exchangers [45, 49–51].

The results of Remenárová et al. [52] on the removal of radionuclide from solution have shown that maximum sorption capacity is achieved though high values indicating that the physical adsorption is not the main binding mechanism. So far many studies [45, 53–56] based on the chemical nature of zeolites stated that ion exchange is the dominating mechanism of metal cation sorption in which cations in the zeolite framework are replaced by metal cations present in water.

Some author modified the surface properties of zeolite in order to enhance the sorption capacity for the removal of radionuclides. For example, Singh and Dutta [57] modified the surface of the zeolite Y type using *n*-octadecyltrichlorosilane for the extraction of ^{137}Cs and ^{90}Sr from aqueous to organic phases [35, 57].

The removal of radionuclides such as ^{137}Cs and ^{134}Ba onto zeolites was investigated by Chmielewská-Horváthová [58]. The examinations were carried out on synthetic aqueous solutions using clinoptilolite and mordenite. In this process, the predominant reaction was also related to the ion exchange of cations that occur in mineral structure, i.e., Ca^{2+}, Mg^{2+}, K^+ and Na^+. Results have also shown that pH of tested solutions is also important in the sorption capacity of tested zeolites, where the best pH values of sorption are almost neutral.

In addition, Elizondo et al. [59] studied the effects of pH solution and particle size on the removal of ^{137}Cs and ^{90}Sr by natural zeolite (clinoptilolite) from liquid radioactive wastes and observed that natural zeolite is an effective filter for the radionuclides from a liquid radioactive waste solution with a pH value of 8.

Consequently, Faghihian [60] used modified zeolites for the removal of radioactive iodine (^{121}I and ^{131}I) with different quaternary alkylammonium ions to replace the exchangeable cations from the zeolite surface.

3.3. Results of radionuclide sorption

The third experiment on radionuclides for the removal of (^{226}Ra and ^{228}Ra) from mine water taken from a hard coal mine KWK Piast was carried out by Chałupnik et al. [61]. The effectiveness of tested zeolite (Na-P1) with regards to the removal of radionuclides was measured in an experiment performed under dynamic conditions.

The effectiveness of zeolite sorbent with regards to the removal radionuclides was measured under dynamic conditions wherein the used mine water (KWK Piast, Poland horizon 650 m) contained the following concentrations of isotopes: ^{226}Ra 4.216 ± 0.534 [$kBq{\cdot}m^{-3}$], ^{228}Ra 7.81 ± 1.46 [$kBq{\cdot}m^{-3}$] (thermal conductivity was 119,000 $\mu S{\cdot}cm^{-1}$ and salinity was 110 $g{\cdot}dm^{-3}$). In this experiment, the column of about 30 cm in diameter was filled with 26 kg of sand and 2 kg of tested zeolite.

The results obtained by Chałupnik et al. [61] on the removal radionuclides (^{226}Ra and ^{228}Ra) have shown high efficiency of up to 98% of the tested zeolites. After passing 120 dm^3 of mine water through the zeolite bed in the column, the concentration of ^{226}Ra decreased to a value of 0.017 $kBq{\cdot}m^{-3}$, and in the case of isotope ^{228}Ra decreased to a value of 0.05 $kBq{\cdot}m^{-3}$. It has been concluded that despite the adsorption of about 500 Bq of ^{226}Ra and about 1000 Bq of ^{228}Ra still no breakthrough of zeolite bed had occurred. They also carried out preliminary laboratory tests on "synthetic waters" and found that for waters with a similar concentration of radium ions as in the experiment a column breakthrough could only occur after passing approximately 1200 dm^3 of water through it with similar characteristics as in the present experiment [61].

The analysis of the water samples obtained after the process of removing radium ions showed that using Na-P1 zeolite for the removal of radionuclides has no effect on the level of calcium or magnesium concentration in the purified water where their concentrations before and after the experiment are similar. The same tested zeolite also did not cause significant changes in the concentration of monovalent metals (sodium and potassium) as well as iron, strontium and manganium in the solution. Moreover, no change was observed in the concentrations of anions (chloride, sulphate, or bicarbonate ions), which are not adsorbed by Na-P1 that is related to the negative charge on their surface [61].

Other authors carried out the removal of radiounuclides using different zeolites among others natural zeolites. Gallant and Prakash [62] carried out dynamic experiment onto the removal of radionuclides such as ^{137}Cs, ^{60}Co and ^{90}Sr from water effluents also rich in Cu, Cd and Zn from nuclear operations using clinoptilolite and chabazite. Experiments have shown that in multication solution caesium is easily removed by both zeolites, so it would be economical to remove caesium first from such solutions using low-cost zeolites [62].

In 2014, Remenárová et al. [45] used synthetic zeolites from fly ash for the removal of caesium from the solution of CsCl under static conditions. They have examined the influence of pH on the sorption of radionuclides and the results have shown that the better sorption was found at pH values ranging from 4 to 5. Nevertheless, the carried out experiments have shown that synthesized zeolite materials could be suitable sorbents for the removal of radiocaesium from liquid radioactive wastes. Moreover, Moattar and Hayeripour [46] have examined natural zeolite (clinoptilolite) for the removal of ^{137}Cs, ^{60}Co and ^{54}Mn from low-level radioactive liquid waste under static and dynamic conditions. In such experiments, zeolites showed the same promising results in the removal of caesium.

El-Naggar et al. [49] used A type synthesized zeolite and evaluated it as an inorganic ion-exchange material for the removal of ^{137}Cs and ^{90}Sr ions from aqueous solutions in both batch and fixed bed column operations. The results have shown that the total ion uptake of these metals and the overall bed capacity decreased by increasing the flow rate and increased by increasing the initial ion concentrations and bed depth. The removal of caesium was also investigated by the International Atomic Energy Agency [63] where tests of the removal of ^{137}Cs ions for aqueous solution were carried out using synthetic NaA-X zeolite. This synthetic zeolite showed a promising sorption behaviour.

Soelberg et al. [64] impregnated zeolite of silver in order to enhance the uptake of volatile radioactive iodine and krypton. Silver impregnated zeolites showed high iodine decontamination factors as high as 105, iodine loading capacities and other adsorption parameters including adsorption rates.

4. Zeolites as petroleum substance sorbents

4.1. Characteristics of petroleum substances

Crude oil belongs to the group of natural hydrocarbon fuels, such as natural gas or coals, which was created as a result of biochemical, chemical and geochemical transformations of organic matter derived from dead plant and animal organisms. It is the liquid mixture of hydrocarbons, oxygen and sulphur compounds, containing trace amounts of various inorganic compounds and elements. The conventional crude oil is mainly composed of hydrogen and carbon in an atomic relation H/C ~ 1.85. Other elements such as sulphur, nitrogen and oxygen are usually not less than 3% (by weight) of its content. The remaining components of the crude oil are usually trace amounts of phosphorus, arsenic, selenium and heavy metals such as vanadium, nickel, iron, cobalt, copper and mercury.

The composition of the hydrocarbons in crude oil consist of paraffinic hydrocarbons (aliphatic), cycloparaffinic (naphthenic) and aromatic among which the cycloparaffins dominate (approximately 50% content). Aliphatic hydrocarbons are the second group of the hydrocarbons present in crude oil in substantial quantities. It is assumed that crude oil comprises *n*-aliphatic (straight) hydrocarbons with the number of carbon atoms from 1 to 60. Except *n*-aliphatic hydrocarbons, branched chain hydrocarbons are present in significant amounts. Their boiling point is lower than the boiling point of unbranched equivalents. Aliphatic hydrocarbons (straight and branched) have lower density and are more volatile than their naphthenic analogues.

Aromatic hydrocarbons contain at least one benzene ring in the molecule. Crude oil usually contains up to 15% of arenes. This type of compounds are typically concentrated in fractions boiling at a higher temperature, such as lubricating oils and residues, where the concentration may exceed 50%. Among the aromatic hydrocarbons, toluene and *m*-xylene are present in the crude oil to the largest quantities. Monocyclic aromatic hydrocarbons constitute a valuable component of the fuel. It should be emphasized that the benzene and its alkyl derivatives (ethyl benzene, toluene and xylene) are compounds with carcinogenic properties and very high toxicity. These features and the relatively good water solubility make them a group of substances that is very hazardous to the environment. On the other hand, polycyclic aromatic hydrocarbons (PAHs) are concentrated in petroleum products such as diesel oils boiling at high temperatures, lubricating oils and residues distillation. They have the ability to accumulate in tissues of living organisms and exhibit the carcinogenic properties [65].

Petroleum distillates are components of commonly used refined products such as gasoline, aviation fuel, diesel fuel and motor oils, hydraulic oils, fuel oils, base oils for lubricants, petroleum waxes, asphalt, petroleum coke and petrochemical raw materials.

Currently, there is a special emphasis on the preparation of such petroleum products that have a minimum negative impact on the environment. Interaction depends on many factors that include the physical, chemical and toxic properties of the product, the manner of its use, scale of consumption and method of disposal. Fuels accounts for about 80% V/V of the processed crude oil and therefore they are the biggest source of oil pollution of the environment [52].

Other products representing approximately 1.5% of processed crude oil pose a smaller risk, albeit not without significance for the environment. In fact, there is still a problem with waste oil and other petroleum waste, which are the substances of potentially dangerous status from the point of view of ecology.

Undesirable effects of petroleum substances are due to their physicochemical properties, including viscosity, density, lipophilicity and hydrocarbon composition that determines these properties.

Among the petroleum compounds particularly dangerous to life and health of living organisms, volatile aromatic hydrocarbons, such as benzene, toluene, ethylbenzene and xylene (BTEX), can be distinguished. These compounds are components of many solvents or fuels, such as petrol or diesel oil [66–68]. They tend to accumulate in groundwater and are also soil or air pollutants [69]. They are carcinogenic, irritate the mucous membranes, cause changes in haematology, central nervous system disorders, respiratory problems, abnormal liver function and kidney function. They act toxic even in small concentrations, and their inhalation or ingestion may result in death [69, 70].

4.2. Sorption mechanisms of petroleum substances on zeolites

Several mechanical, biological and chemical methods are used to liquidate oil spills from water and paved roads [11, 33, 71–76]. Among these methods, one of the most popular is the sorption method using sorbents of different origins. Sorbents are solids with developed surface and able to bond molecules on their surface (from liquid or gaseous phase). These sorption features are characteristics of zeolite minerals.

The most commonly used adsorbents for oil spills on paved surfaces (roads, asphalt and land) are those of mineral origin. For this type of materials, the sorption process takes place according to the physical sorption including two main mechanisms. The first one is a stage of sorption associated with filling of the available pores in the material and the capillary pores between the grains of the sorbent according to mechanism involving capillary (mass transport). Capillarity depends on, among others, the effective diameter of the capillary, or surface energy of the interior wall of the capillary and viscosity of the oil. After this step, sorption occurs on the outer surface and an oily film layer is created around the grain of sorbent (or optionally around the agglomerate grains). On the surface of the sorbent grains, oil can create a uniform layer (film) or irregular aggregates, depending on the morphology of the material surface, its irregularities and the properties of the oil [77].

The adsorption of petroleum compounds from aqueous solutions (e.g., solubilized form of organic compounds such as BTEX) on porous sorbents involves diffusion process, the course of which may include transport of the molecules in the vicinity of the adsorbent, the diffusion in a boundary layer, the diffusion in sorbent's capillary, surface diffusion and intra-particle diffusion.

4.3. Results of petroleum substances sorption

Some relevant papers concerning removal of petroleum substances indicate the possibility of using zeolites for this purpose, including those derived from fly ash.

A comparison of sorption properties in relation to diesel fuel (Biodiesel and Verva ON) of zeolites from fly ash on the zeolite types of structures of Na-P1 and Na-X and natural clinoptilolite as well as on commercial silica-based sorbent (Absodan) was presented by Bandura et al. [78, 79]. They received the sorption capacity of the order of 1.24–1.40 g/g for Na-P1 zeolite, 0.91–1.13 g/g for Na-X zeolite and 0.31 g/g for clinoptilolite. For commercial sorbent, the sorption capacity was 0.83–0.95 g/g (**Figure 1**).

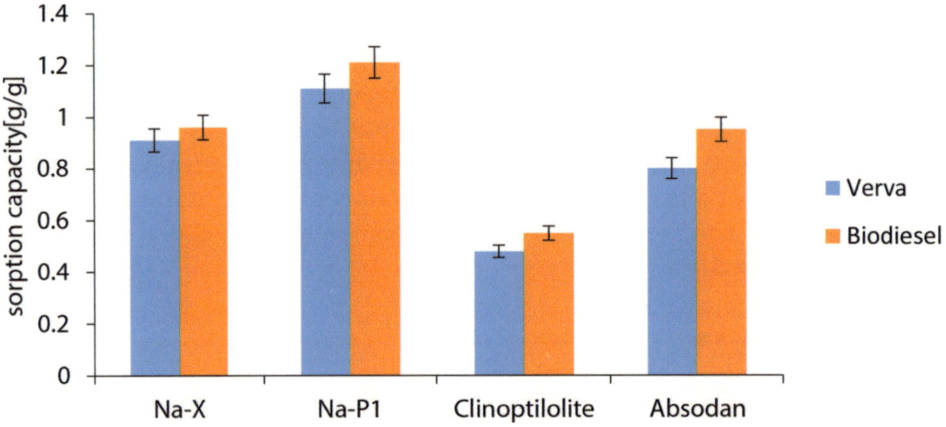

Figure 1. Sorption capacity of zeolite materials.

Na-P1 and Na-X zeolites also showed good sorption properties with respect to oils similar to sepiolite and vermiculite and some commercial sorbents. Importantly, Na-P1 and Na-X zeolites have higher sorption capacity with respect to oil in comparison of zeolites tested in the works of Carmody et al. [77] and Ankowski [80]. The zeolites from the fly ash, i.e., Na-X and Na-P1, can also be used as absorbents of volatile organic compounds [81].

In this direction, Ankowski [80] examined zeolites from fly ash granulated with clay, whose capacity to diesel fuel was 0.51–0.57 g/g. In addition, the capacity of sodium aluminium silicate and commercial sorbents Eco-dry and dry-Lite was determined, which amounted to 0.27–0.43, 1.11, 0.77 g/g, respectively.

Sakthivel et al. [82] synthesized zeolites from fly ash by hydrothermal conversion and fusion methods, and then zeolite products were modified by organic compounds. The sorption capacity of zeolites without modification was 0.37 g/g for zeolite obtained by the hydrothermal method and 1.22 for zeolite obtained by fusion. It is worth noting that the sorption capacity of zeolites modified was in the range of 0.21–1.15 g/g, lower than those that were not modified. These results indicate that organic modification of zeolites is not in all cases profitable with respect to the removal of petroleum substances. It turned out that this type of modification of the sorbent surface blocks the pores of the zeolite, thus reducing its specific surface area and limiting the access of oil into the pores.

The results of this research indicate the possibility of the use of zeolites from fly ash as a mineral sorbent to remove oil spills from paved surfaces.

http://dx.doi.org/10.5772/64362

5. Conclusions

The present results have shown that zeolite can be considered promising materials for the removal of toxic compounds, such as mercury, radionuclides, and petroleum substances.

The results of mercury removal in pure stream of gas have shown that silver impregnation greatly improves zeolites efficiencies. The highest uptake achieved Na-X Ag-impregnated that adsorbed significantly more mercury than the commercial available activated carbon with bromine. For the removal of mercury in flue gas only Na-X Ag-impregnated was tested (powder and granulate). The tested zeolite reduced the level of mercury in the flue gas where the long-time mercury breakthrough ranges from 15 to 40% depending on the experimental conditions. The better results were obtained for granulated material.

Results of removing radium isotopes indicate high efficiency of the process at a level of 98% which indicates a high application potential of synthetic zeolites for the purification of mine water from radionuclides. Other radionuclides, mainly caesium and strontium, tested by other authors are also removed perfectly from various solutions using zeolites (natural and synthetic).

Results have shown that the sorption of diesel fuels and used oil on clinoptilolite, Na-P1 and Na-X, and commercial sorbent mainly constitutes physical characters. Synthetic zeolites have around twice higher sorption capacity for oils than commercial sorbents. Higher sorption capacities were achieved for oils of higher densities.

Acknowledgements

This work was financed by the National Centre for Research and Development (NCBR) under contract number LIDER/384/L-6/14/NCBR/2015 and statutory fund no. S12/II/B/2016.

Author details

Magdalena Wdowin[1*] and Wojciech Franus[2]

*Address all correspondence to: wdowin@meeri.pl

1 Mineral and Energy Economy Research Institute, Polish Academy of Sciences, Kraków, Poland

2 Civil Engineering and Architecture Faculty, Lublin University of Technology, Lublin, Poland

References

[1] Plant JA, Korre A, Reeder S, Smith B, Voulvoulis N. Chemicals in the environment: implications for global sustainability. Appl. Earth Sci. (Trans. Inst. Min. Metall. B). 2005; 114(B65): 1–33.

[2] RCEP. 24th report: chemicals in products: safeguarding the environment and human health. ISBN 0 10 158272 2. 2003.

[3] EAA/UNEP (European Environment Agency and United Nations Environment Programme). 1999. Chemicals in the European Environment: Low Doses, High Stakes? The EEA and UNEP Annual Message 2 on the State of Europe's Environment.

[4] Wdowin M, Franus M, Panek R, Bandura L, Franus W. The conversion technology of fly ash into zeolites. Clean Technol. Environ. Policy. 2014; 16(6): 1217–1223.

[5] Mumpton FA. La roca magica: uses of natural zeolites in agriculture and industry. 1985; 3463–3470. doi: 10.1073/pnas.96.7.3463.

[6] Wdowin W, Macherzyński M, Panek R, Górecki J, Franus W. Investigation of the mercury vapour sorption from exhaust gas by an Ag-X zeolite. Clay Mineral. 2015; 50(1): 31–40.

[7] Pacyna EG, Pacyna JM, Steenhuisen F, Wilson SJ. Global anthropogenic mercury emission inventory for 2000. Atmos. Environ. 2006; 40: 4048–4063.

[8] United Nations Environment Programme (UNEP). Global Mercury Assessment 2013: Sources, Emissions, Releases and Environmental Transport, Geneva, Switzerland. 2013.

[9] Srivastava RK, Hutson N, Martin B, Princiotta F, Staudt J. Control of mercury emissions from coal-fired electric utility boilers. Environ. Sci. Technol. 2006; 40(5): 1385–1393.

[10] Gustin MS, Lindberg SE, Weisberg PJ. An update on the natural sources and sinks of atmospheric mercury. Appl. Geochem. 2008; 23: 482–493.

[11] Zytner RG. Sorption of benzene, toluene, ethylbenzene and xylenes to various media. J. Hazard. Mater. 1994; 38: 113–126.

[12] Lopez-Anton MA, Tascon JMD, Martinez-Tarazona MR. Retention of mercury in activated carbons in coal combustion and gasification flue gases. Fuel Process. Technol. 2002; 77–78: 353–358.

[13] Schroeder WH, Munthe J. Atmospheric mercury—an overview. Atmos. Environ. 1998; 32: 809–822.

[14] Senior CL, Sarofim AF, Zeng T, Helble JJ, Mamani-Paco R. Gas-phase transformations of mercury in coal-fired power plants. Fuel Process. Technol. 2000; 63: 197–213.

[15] Galbreath KC, Zygarlicke ChJ. Mercury Transformation in coal combustion flue gas. Fuel Process. Technol. 2000; 65–66: 289–310.

[16] Morency JR, Panagiotou T, Senior CL. Laboratory Duct Injection of a Zeolite-Based Mercury Sorbent, Presented at the Annual Meeting of the Air & Waste Management Association, Salt Lake City, UT, June 2000.

[17] Wdowin M, Wiatros-Motyka MM, Panek R, Stevens LA, Franus W, Snape CE. Experimental study of mercury removal from exhaust gases. Fuel. 2014; 128: 451–457. doi: 10.1016/j.fuel.2014.03.041.

[18] Agarwal H, Stenger HG. Development of a predictive kinetic model for homogeneous Hg oxidation data. Mat. Comput. Modell. 2007; 45(1–2): 109–125.

[19] Sloss L, Smith M. Trace element emissions. IEA Coal Research. Report CCC/34, IEA Coal Research, London, United Kingdom. 2000, 83 pp.

[20] Wang J, Clements B, Zanganeh K. An interpretation of flu-gas mercury speciation data from a kinetic point of view. Fuel. 2003; 82(8): 1009–1011.

[21] Xu M, Qiao Y, Zheng C, Li L, Liu J. Modeling of homogenous mercury speciation using detailed chemical kinetics. Combust. Flame. 2003; 132(1–2): 208–218.

[22] Presto AA, Granite EJ. Survey of catalysts for oxidation of mercury in flue gas. Environ. Sci. Technol. 2006; 40(18): 5601–5609.

[23] Jurng J, Lee TG, Lee GW, Lee SJ, Kim BH, Seier J. Mercury removal from incineration flue gas by organic and inorganic adsorbents. Chemosphere. 2002; 47: 907–913.

[24] Abrams L, Corbin DR. Probing inter-zeolite space, w: inclusion chemistry with zeolites; nanoscale materials by design, Herron N. I, Corbin D.R. [red.], Kluwer, Dordrecht. 1995.

[25] Barrer RM, Whiteman JI. Mercury Uptake in various cationic forms of several zeolites. J. Chem. Society A. 1967; 20: 19–25.

[26] Dong J, Xu Z, Kuznicki SM. Mercury removal from flue gases by novel regenerable magnetic nanocomposite sorbents. Environ. Sci. Technol. 2009; 43: 3266–3271.

[27] Wei Z, Luo Y, Li B, Cheng Z, Wang J, Ye Q. Microwave assisted catalytic removal of elemental mercury from flue gas using Mn/zeolite catalyst. Atmos. Pollut. Res. 2015; 6: 45–51.

[28] Wiatros-Motyka MM, Sun Ch-g., Stevens LA, Snape CE. High capacity co-precipitated manganese oxides sorbents for oxidative 4 mercury capture. Fuel. 2013; 109: 559–562.

[29] Qu Z, Yan NQ, Liu P, Chi YP, Jia J. Bromine chloride as an oxidant to improve elemental mercury removal from coal–fired flue gas. Environ. Sci. Technol. 2009; 43: 8610–8615.

[30] Li JR, He C, Shang XS, Chen JS, Yu XW, Yao YJ. Oxidation efficiency of elemental mercury in flue gas by SCR De–NOX catalysts. J. Fuel. Chem. Technol. 2012; 40: 241–246.

[31] Baltrus J, Granite E, Pennline H, Stanko D, Hamilton H, Rowsell L et al. Surface characterization of palladium–alumina sorbents for high-temperature capture of mercury and arsenic from fuel gas. Fuel. 2009; 89: 6, 1323–1325.

[32] Dunham GE, De Wall RA, Senior CL. Fixed-bed studies of the interaction between mercury and coal combustion fly ash. Fuel Process. Technol. 2003; 82: 197–213.

[33] Mathur AK, Majumder CB, Chatterjee S. Combined removal of BTEX in air stream by using mixture of sugar cane bagasse, compost and GAC as biofilter media. J. Hazard. Mater. 2007; 148: 64–74. doi:10.1016/j.jhazmat.2007.02.030.

[34] Qi H, Xu W, Wang J, Tong L, Zhu T. Hg^0 removal from flue gas over different zeolites modified by $FeCl_3$. J. Environ. Sci. 2015; 28: 110–117.

[35] Yıldız B, Erten HN, Kıs M. The sorption behavior of CS+ ion on clay minerals and zeolite in radioactive waste management: sorption kinetics and thermodynamics. J. Radioanal. Nucl. Chem. 2011; 288: 475–483.

[36] Couling DJ, Nguyen HV, Green WH. Screening of metal oxides and metal sulfides as sorbents for elemental mercury at elevated temperatures. Fuel. 2012; 97: 783–795.

[37] Presto AA, Granite EJ. Impact of sulphur oxides on mercury capture by activated carbon. Environ. Sci. Technol. 2007; 41: 6579–6584.

[38] Liu Y, Bisson TM, Yang H, Xu Z. Recent developments in novel sorbents for flue gas clean up. Fuel Process. Technol. 2010; 91: 1175–1197.

[39] Moutsatsou A, Stamatakis E, Hatzitzotzia K, Protonotarios V. The utilization of Ca-rich and Ca–Si-rich fly ashes in zeolites production. Fuel. 2006; 85: 657–663.

[40] His H-Ch, Mark J, Rood MJ, M.ASCE., Rostam-Abadi M, Chen S, Chang R. Mercury Adsorption properties of sulfur-impregnated adsorbents. J. Environ. Eng. 2000; 128(11): 1080–1089.

[41] Morency JR, Panagiotou T, Lobo RF. Control of mercury emissions in utility power plants. EPRI-DOE-EPA Combined Utility Air Pollutant Control Symposium ("The Mega Symposium") 16–20 August 1999, Atlanta, GA Copyright © 1999 Electric Power Research Institute.

[42] Morency JR, Panagiotou T, Senior CL. Zeolite sorbent that effectively removes mercury from flue gases. Filtration Sep. 2002; 39(7): 24–26.

[43] Pavlish JH, Sondreal EA, Mann MD, Olson ES, Galbreath KC, Laudal DL, Benson SA. Status review of mercury control options for coal-fired power plants. Fuel Process. Technol. 2003; 82: 89–165.

[44] Khan SA. Sorption of the long-lived radionuclides cesium-134, strontium-85 and cobalt-60 on bentonite. J. Radioanal. Nucl. Chem. 2003; 258: 3–6.

[45] Remenárová L, Pipíška M, Florková E, Augustín J, Rozložník M, Hostin S, Horník M. Radiocesium adsorption by zeolitic materials synthesized from coal fly ash. Nova Biotechnol. Chim. 2014; 13(1): 57–72.

[46] Moattar F, Hayeripour S. Application of chitin and zeolite adsorbents for treatment of low level radioactive liquid wastes. Int. J. Environ. Sci. Technol. 2004; 1(1): 45–50.

[47] Pociask-Karteczka J, Jasińska M, Mietelski JW. The influence of radionuclides released by Silesia coal mine activity on natural environment in the Upper Vistula Basin (Poland). IAHS Publ. 1997; 243: 103–108.

[48] Abd El-Rahman KM, El-Kamash A, El-Sourougy MR, Abdelmoniem NM. Thermodynamic modeling for the removal of Cs^{+}, Sr^{2+}, Ca^{2+}, and Mg^{2+} ions from aqueous waste solutions using zeolite A. J. Radioanal. Nucl. Chem. 2006; 268: 221–230.

[49] El-Naggar MR, El-Kamash AM, El-Dessouky MI, Ghonaim AK. Two-step method for preparation of Na A-X zeolite blend from fly ash for removal of cesium ions. J. Hazard. Mater. 2008; 154: 963–972.

[50] Goni S, Guerrero A, Lorenzo MP. Efficiency of fly ash belite cement and zeolite matrices for immobilizing cesium. J. Hazard. Mater. 2006; B137: 1608–1617.

[51] Li J, Gong ZW, Jianlong W. Solidification of low-level radioactive resins in ASC zeolite blends. Nucl. Eng. Design. 2005; 235: 817–821.

[52] Rutkowska J, Surygała J, Joanna K. (not published) Vademecum rafinera–Ropa naftowa: właściwości, przetwarzanie, produkty (Vademecum refiner-Oil: properties, processing, products), Warszawa, 2006: Wydawnictwa Naukowo-Techniczne; 280 pp. (in polish)

[53] Erdem E, Karapinar N, Donat R. The removal of heavy metal cations by natural zeolites. J. Colloid. Interf. Sci. 2004; 280: 309–314.

[54] Merrikhpour H, Jalali M. Comparative and competitive adsorption of cadmium, copper, nickel, and lead ions by Iranian natural zeolite. Clean Technol. Environ. Policy. 2013; 15: 303–316.

[55] Qiu W, Zheng Y. Removal of lead, copper, nickel, cobalt, and zinc from water by a cancrinite-type zeolite synthesized from fly ash. Chem. Eng. J. 2009; 145: 483–488.

[56] Sprynskyy M, Buszewski B, Terzyk AP, Namieśnik J. Study of the selection mechanism of heavy metal (Pb^{2+}, Cu^{2+}, Ni^{2+}, and Cd^{2+}) adsorption on clinoptilolite. J. Colloid. Interf. Sci. 2006; 304: 21–28.

[57] Singh R, Dutta PK. Use of surface-modified zeolite Y for extraction of metal ions from aqueous to organic phase. Micropor. Mesopor. Mater. 1999; 32: 29–35.

[58] Chmielewska-Horvfithovfi E. The interaction mechanisms between aqueous solutions of 137Cs and 134Ba radionuclides and local natural zeolites for an deactivation scenario. J. Radioanal. Nucl. Chem. 1998; 227(1–2): 151–155.

[59] Elizondo NV, Ballesteros E, Kharisov BI. Cleaning of liquid radioactive wastes using natural zeolites. Appl. Radiat. Isot. 2000; 52: 27–30.

[60] Faghihian H, Akbar Malekpour A, Maragheh MG. Removal of radioactive iodide by surfactant-modified zeolites. Adsorption Sci. Technol. 2013; 21(4): 373–381.

[61] Chałupnik S, Franus W, Wysocka M, Gzyl G. Application of zeolites for radium removal from mine water. Environ. Sci. Pollut. Res. 2013; 20(11): 7900–7906.

[62] Gallant J, Prakash A. Removal of Selected Radionuclides and Metal Contaminants by Natural Zeolites from Liquid Effluents, The Canadian Nuclear Society 30th Annual CNS Conference and 33rd CNS/CAN Student Conference 2009.

[63] International Atomic Energy Agency. Application of Ion Exchange Processes for the Treatment of Radioactive Waste and Management of Spent Ion Exchangers. TRS No. 408; IAEA: Vienna, Austria. 2002.

[64] Soelberg NR, Garn TG, Greenhalgh MR, Law JD, Jubin R, Strachan DM, Thallapally PK. Radioactive iodine and krypton control for nuclear fuel reprocessing facilities. *Sci. Technol. Nucl.* Ins. 2013. http://dx.doi.org/10.1155/2013/702496.

[65] Franci CD, Guillemette M, Pelletier E, Chastel O, Bonnefoi S, Verreault J. Endocrine status of a migratory bird potentially exposed to the Deepwater Horizon oil spill: a case study of northern gannets breeding on Bonaventure Island, Eastern Canada. Sci. Total Environ. 2014; 473–474: 110–116. doi:10.1016/j.scitotenv.2013.12.006.

[66] Aivalioti M, Pothoulaki D, Papoulias P, Gidarakos E. Removal of BTEX, MTBE and TAME from aqueous solutions by adsorption onto raw and thermally treated lignite. J. Hazard. Mater. 2012; 207–208: 136–146. doi:10.1016/j.jhazmat.2011.04.084.

[67] Nourmoradi H, Nikaeen M, Khiadani (Hajian) M. Removal of benzene, toluene, ethylbenzene and xylene (BTEX) from aqueous solutions by montmorillonite modified with nonionic surfactant: equilibrium, kinetic and thermodynamic study. Chem. Eng. J. 2012; 191: 341–348. doi:10.1016/j.cej.2012.03.029.

[68] Su F, Lu C, Hu S. Adsorption of benzene, toluene, ethylbenzene and p-xylene by NaOCl-oxidized carbon nanotubes. Colloids Surf. A Physicochem. Eng. Asp. 2010; 353: 83–91. doi:10.1016/j.colsurfa.2009.10.025.

[69] Vidal CB, Raulino GSC, Barros AL, Lima ACA, Ribeiro JP, Pires MJR, et al. BTEX removal from aqueous solutions by HDTMA-modified Y zeolite. J. Environ. Manage. 2012; 112: 178–185. doi:10.1016/j.jenvman.2012.07.026.

[70] Moura CP, Vidal CB, Barros AL, Costa LS, Vasconcellos LCG, Dias FS, et al. Adsorption of BTX (benzene, toluene, o-xylene, and p-xylene) from aqueous solutions by modified periodic mesoporous organosilica. J. Colloid. Interf. Sci. 2011; 363: 626–634. doi:10.1016/j.jcis.2011.07.054.

[71] Broje V, Keller AA. Improved mechanical oil spill recovery using an optimized geometry for the skimmer surface. Environ. Sci. Technol. 2006; 40: 7914–7918.

[72] Chagas-Spinelli ACO, Kato MT, de Lima ES, Gavazza S. Bioremediation of a tropical clay soil contaminated with diesel oil. J. Environ. Manage. 2012; 113: 510–516. doi: 10.1016/j.jenvman.2012.05.027.

[73] Ji F, Li C, Dong X, Li Y, Wang D. Separation of oil from oily wastewater by sorption and coalescence technique using ethanol grafted polyacrylonitrile. J. Hazard. Mater. 2009; 164: 1346–1351. doi:10.1016/j.jhazmat.2008.09.048.

[74] Nikolajsen K, Kiwi-Minsker L, Renken A. Structured Fixed-bed adsorber based on zeolite/sintered metal fibre for low concentration VOC removal. Chem. Eng. Res. Des. 2006; 84: 562–568. doi:10.1205/cherd.05220.

[75] Półka M. Analiza skuteczności wybranych sorbentów wykorzystanych do pochłania-nia par substancji ropopochodnych działaniach ratowniczych (Efficiency analysis of the sorbents used to adsorb the vapors of petroleum products during rescue and firefighting actions). Przem Chem. Chem. 2015; 1: 109–113. doi: 20 10.15199/62.2015.3.13 (in polish)

[76] Saikia RR, Deka S. Removal of hydrocarbon from refinery tank bottom sludge employ-ing microbial culture. Environ. Sci. Pollut. Res. Int. 2013; 20: 9026–9033. doi:10.1007/s11356-013-1888-2.

[77] Carmody O, Frost R, Xi Y, Kokot S. Surface characterisation of selected sorbent materials for common hydrocarbon fuels. Surf. Sci. 2007; 601: 2066–2076.

[78] Bandura L, Franus M, Józefaciuk G, Franus W. Synthetic zeolites from fly ash as effective mineral sorbents for land-based petroleum spills cleanup. Fuel 2015; 147: 100–107. doi:10.1016/j.fuel.2015.01.067.

[79] Bandura L, Franus M, Panek R, Woszuk A, Franus W. Charakterystyka zeolitów i ich zastosowanie jako adsorbentów substancji ropopochodnych (Characterization of zeolites and their use as adsorbents of petroleum substances). Przem. Chem. 2015; 94:323–327 (in polish).

[80] Ankowski A. Wykorzystanie zeolitów z popiołów lotnych w sorpcji substancji ropopochodnych w warunkach rzeczywistych (The use of zeolites from fly ash in the sorption of oil products in real conditions). Politech. Częstochowska. 2009, 11p. (in polish).

[81] Bandura L, Panek R, Rotko M, Franus W. Synthetic zeolites from fly ash for an effective trapping of BTX in gas stream. Micropor. Mesopor. Mater. 2016; 223: 1–9.

[82] Sakthivel T, Reid DL, Goldstein I, Hench L, Seal S. Hydrophobic high surface area zeolites derived from fly ash for oil spill remediation. Environ. Sci. Technol. 2013; 47: 5843–5850.

Section 2

Natural and Synthetic Zeolites: Applications

Chapter 5

Industrial Zeolite Molecular Sieves

Kristin Gleichmann, Baldur Unger and Alfons Brandt

Additional information is available at the end of the chapter

http://dx.doi.org/10.5772/63442

Abstract

Zeolite molecular sieves are used in industrial applications since more than 60 years, mainly as highly efficient adsorbents for separation processes in gas or liquid phase. Zeolite molecular sieves may be applied in powder form, preferably in static applications, but to a much larger extent as shaped material in both static and dynamic (flowing media) applications. Many shaping technologies for molecular sieves have been developed over the last decades, reflecting the different requirements for molecular sieves in different applications. This review deals with the influence of the applied zeolite synthesis and shaping methods for hydrophilic zeolite molecular sieves (procedures, materials, recipes) on the potential industrial applications; thereby considering powders, binder-containing shapes as well as binderless shapes (including compact structures such as honeycombs, multi-channel, and foam-like structures). Due to new challenges from the market, more specialized, tailor-made types of zeolite molecular sieves are required. Such a higher specialization can be achieved by applying new types of zeolites or zeolite-like materials, modified synthesis and/or post synthesis treatments, and by modified, to the needs of the application adjusted shaping processes.

Keywords: zeolite, molecular sieve, adsorption, fixed bed, compact zeolite bodies

1. Introduction

Zeolite molecular sieves are used in industrial applications since more than 60 years, mainly as highly efficient adsorbents for separation processes in gas or liquid phase. The current market volume for molecular sieves (especially for zeolites for adsorption applications) is 150.000–200.000 t with a sales volume of 450–550 Mio. USD with an annual growth forecast up to 2020 of 3–5% [1–3]. Growth drivers are environmental protection and conservation of resources (deeper processing of crude oil, using of coal and natural gas (especially from off-shore and shale gas deposits coming [4–6]) as raw material for petrochemical products in the so-called coal-to-liquid (CTL) or gas-to-liquid (GTL) processes) [7–9].

Zeolite molecular sieves can be applied in powder form, preferably in static applications. During the polyurethane production, zeolite 3A powder is mixed with the polyol component to enable exclusion of moisture and prevent the chemical reaction of water with the isocyanate component [10]. To support the material handling, and to avoid, for example, dust formation molecular sieve beads are used in non-regenerative applications, too. Dry sealing plug or small dry bags filled with zeolite A (or silica) [11, 12] to guaranty a dry transport or storage of humidity sensitive goods such as electronical products, leather ware, or pharmaceuticals are common [11, 12]. The application of zeolite 3A in insulating glass panes is well-known, too. By putting the adsorbent material as desiccant in the perforated spacer between the two or three window panes moisture condensation between the panes (fogging) is prevented [13]. Refrigerant drying [14, 15] is also a static adsorption process, where in the coolant circuit formed, water and acidic components are removed [14, 15].

Besides the mentioned application examples, shaped zeolite molecular sieve material in flowing media is used to a much larger extent. For instance, the purification of technical hydrogen (removal of nitrogen and carbon monoxide) on zeolite 5A [16], the pre-purification of raw air on zeolite NaX [17] for the cryogenic air separation or the adsorptive oxygen enrichment, for example, on zeolite type 5A [18] or LiX [19, 20] should be mentioned in this place. Applied technologies for purification and drying of natural gas are operating with zeolite molecular sieves [21] of zeolite 5A and/or NaX type [22]. In petro chemistry applications especially in steam cracking plants, light hydrocarbons are separated by low temperature rectification. To avoid ice crystal and gas hydrate formation, cracked gases have to be dried to a residual water content <1 ppm [23]. Zeolite molecular sieves of type 3A meet the requirements for that complex process [22]. For the sake of completeness, the principle of the simulating moving bed (SMB) should also be mentioned. This practically continuous liquid phase adsorption process is based on the countercurrent extraction principle [24]. The so-called Molex process, where paraffin mixtures (zeolite molecular sieve of type 5A [25, 26]) are separated, and the so-called Parex processes, where *p*-xylen is separated from C_8 aromatic mixture (zeolite molecular sieve of type KY or BaX [27]) are known SMB processes. Furthermore, the dehydration of preferably bio-ethanol above the azeotropic point to more than 99.5% on zeolite 3A is common practice [28, 29]. Besides the mentioned examples, the use of zeolite molecular sieves in automotive applications such as drying of brake air in rail vehicles, trucks, or busses is also a well-known application. To avoid corrosion or freezing caused by humidity, such pneumatically working breaking systems are equipped with zeolite 4A molecular sieve or silica gel-containing desiccant cartridges [22].

Another field for the application of zeolite molecular sieves is the utilization of the adsorption heat during the adsorption process. Different applications such as building heating [30–32] or enhanced household devices such as dish washers [33, 34] based on the thermo-chemical energy properties of zeolites are introduced into the market.

The above-mentioned examples clearly demonstrate the demand for zeolite molecular sieves with special properties depending on the different requirements in the application.

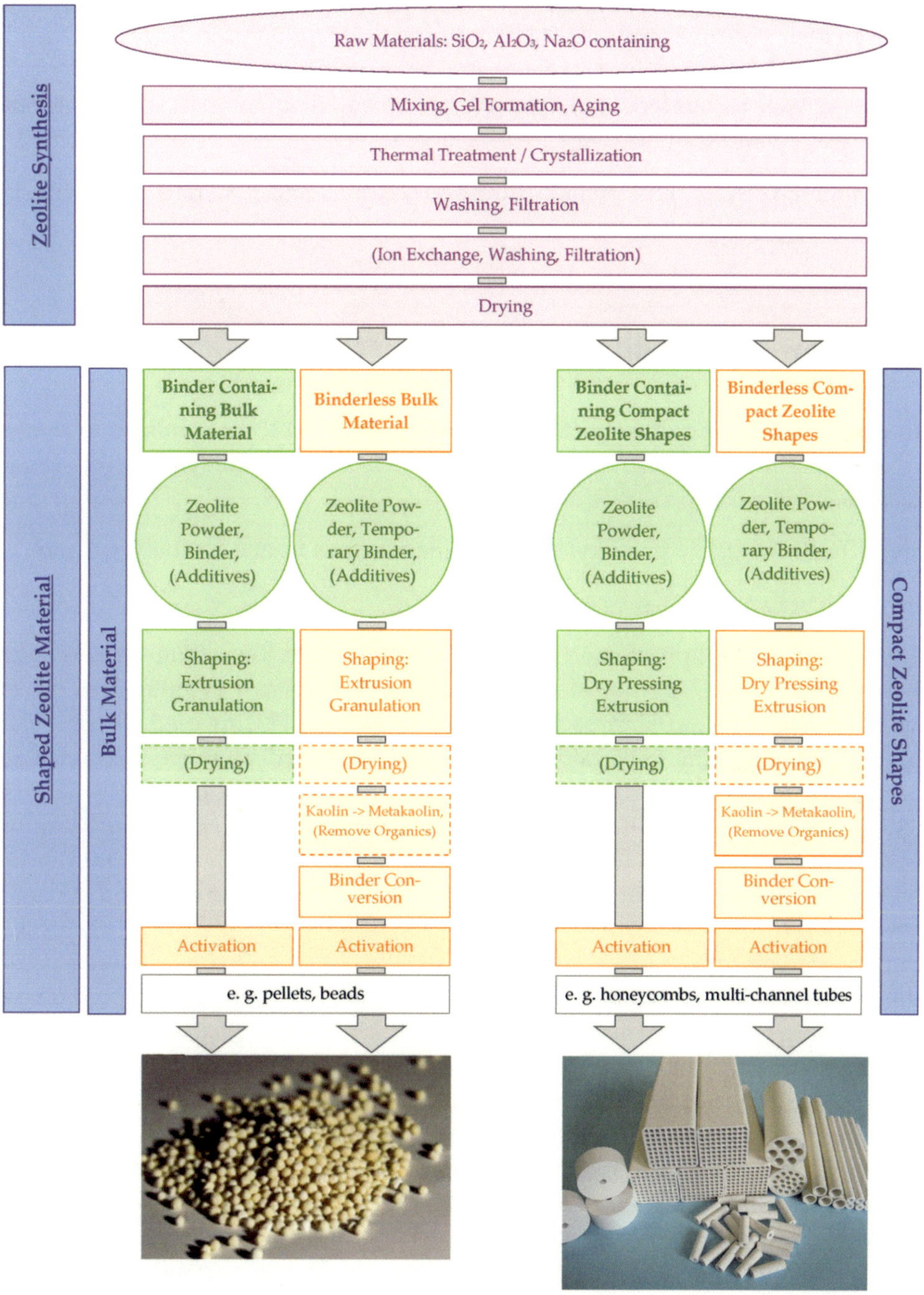

Figure 1. Overview of the manufacturing of different zeolite molecular sieves (figure compact zeolite bodies [35]).

Many synthesis and shaping technologies for molecular sieves were developed over the last decades. Within the following descriptions, the influence of the applied synthesis and shaping methods for hydrophilic zeolite molecular sieves of zeolite A, X, and NaY on their properties, and thus, on their potential applicability in industrial adsorption processes is discussed. The procedures, materials, and recipes for the manufacturing of zeolite powders, binder-containing shapes as well as binderless shapes are described. Also compact structures such as honeycombs and multi-channel tubes are considered.

Figure 1 summarizes different manufacturing procedures and materials dealt with in this review.

2. Zeolite powder with different particle sizes

Within the manufacturing of synthetic zeolite molecular sieves, the synthesis of the related zeolite powder is first step. There are two different principles for the synthesis of aluminum-rich zeolites in industrial processes:

- Manufacturing of highly pure crystalline zeolite powders from a synthesis gel and
- Conversion of natural minerals into zeolite matter.

Usually, in the industrial production of zeolite A, X, and NaY, the sodium form is manufactured (except the zeolite LSX synthesis [36]). Therefore, different raw materials as silicon source such as silicic acid, colloidal silica, sodium silicate, or silica gel [37] are used. Possible alumina sources are alumina trihydrate, sodium aluminate, activated alumina, gamma alumina. Sodium may be obtained from sodium hydroxide which controls simultaneously the pH value [37]. A convenient way to prepare such a reactant mixture is to pour under more or less intensive stirring an aqueous solution containing the aluminum source and sodium hydroxide into an aqueous solution containing the silica source. After gel formation and generating a homogeneous mixture, the mixture is heated up to maximal 120°C at atmospheric pressure or moderate overpressure [37]. After crystallization, the zeolite suspension is filtered at temperatures between room temperature and reaction temperature [37], for example, in a filter press [38] or a centrifuge [39, 40] or a vacuum belt filter [40]. The zeolite crystals are washed directly on the filter device till a pH value between 9 and 12 [37] is achieved. If desired, the zeolite powder in its (virgin) sodium form can be ion exchanged or otherwise modified. For example, the filter cake can be suspended in a suitable salt solution and stirred at a suitable temperature. Alternatively, the exchange step is done in a column [41] or directly on the filter press [42] with a suitable salt solution at a suitable temperature. After the ion-exchange procedure is completed, the zeolite is washed as described above. Afterwards, the filter cake is dried in a direct or indirect heated thermal treatment process, as, for example, spray drying, spin flash drying, or long gap mill drying [40, 43]. The dried zeolite powder is processed in the following steps. The mother liquor from the filtration step can usually (at least to a certain extent) be reused in the following zeolite synthesis batch [37, 44].

The manufacturing of zeolites by conversion of minerals is not really a commonly used process. In several papers which deal with the synthesis of zeolite A, X, and NaY from minerals, kaolin or metakaolin is used as aluminum source [45–47]. A disadvantage of such procedures is the presence of impurities in the applied minerals [45], but some advantages (simple treatment of the mother liquor and a nearly waste-free production) support the application of that synthesis method especially in the case of zeolite NaY [46].

The properties of zeolite molecular sieves can be influenced by the properties of the starting zeolite powder. For instance, the hydrothermal stability or desorption characteristics of FAU vary with the chemical composition, especially the aluminum content [48]. The particle size, particle size distribution, and the surface characteristics of single zeolite crystals affect the adsorption properties as well as the mechanical stability of zeolite molecular sieves. By varying the synthesis gel composition, time, or temperature or by adding additives to the synthesis gel, the mentioned properties of the zeolite powder can be changed [49, 50]. Depending on the intended applications, the particle size distribution is adjusted. Besides the described classical gel synthesis [51] resulting in particles of about 5 μm, microparticles (particle size of approx. 1 μm) and nanoparticles (particle sizes between 50 and 100 nm) can be achieved via new synthesis routes [52].

3. Binder-containing zeolite molecular sieves

In dynamic adsorption processes, where the adsorbent has to be regenerated frequently, the fluid dynamics (e.g., the pressure drop over the fixed bed of molecular sieves) has to be taken into account. Therefore, the zeolite has to be applied in a suitable particle size (usually in the form of shaped mechanically stable macroscopic particles such as beads/spheres or extrudates/cylinders).

Because pure zeolite powder does not exhibit binding properties, an appropriate binder needs to be applied to form the above-mentioned macroscopic particles. **Figure 2** shows the main principles for the manufacturing of conventional binder-containing zeolite molecular sieves (bulk material).

The zeolite component is the main component of the pre-mixture and can be applied as dried zeolite powder (occasionally as filtercake). Usually, the zeolite component is modified (as described above, e.g., ion exchanged or dealuminated) prior to the shaping step. Besides the zeolite component a binder material is put into the pre-mixture. Mineral binders such as bentonite, attapulgite, or kaolin are common [53, 54] as well as synthetic binders such as alumina, silica [54, 55], or a mixture thereof [56]. The type of binder material to be applied depends on the shaping procedure and on the final application of the zeolite molecular sieve. Bentonite (montmorillonite) for example is commonly used for extrusion [57, 58], attapulgite for beading [59]. It should be noted that (natural) mineral binders may contain impurities (e.g., catalytically active heavy metals), which could cause undesired side reactions in the final application. If necessary, additives such as shaping aids (e.g., mineral oil, PVA, starch derivate solutions [60]) are put into the pre-mixture to make the shaping process most effective. Such mainly organic compounds are burnt out in subsequent thermal treatment steps. In order to create a defined secondary pore system (see below) thermally or chemically, removable spacers (e.g., cellulose fibers or soluble salts) are added to the pre-mixture. In most of the bulk zeolite molecular sieves, a binder content of around 20% (anhydrous basis) is used. But also lower (down to around 5%) and higher (above 30%) binder amounts are described [59].

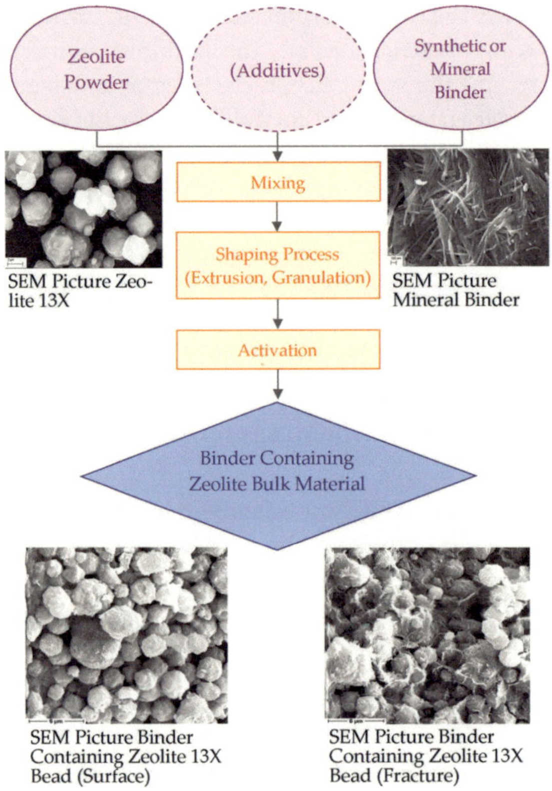

Figure 2. Scheme of the principle manufacturing procedures for binder-containing zeolite molecular sieve shapes (bulk material).

After the (mainly dry) mixing of all aforementioned materials in the pre-mixture usually a certain amount of water is added in order to adjust the viscosity and plasticity of the mixture for the related shaping process that can be agglomeration granulation [59], spray granulation [61], or extrusion [62]. After the shaping process, the molecular sieve has to be dried and activated in a thermal step, for example, in a rotary kiln [63] or belt calciner [64] to remove the water and other adsorbed compounds. The applied temperature is often higher than required for the zeolite activation, since the binder system needs such higher temperatures for setting. The zeolite component in zeolite NaX beads with attapulgite as binder materials, for example, needs not more than 350–400°C for a complete activation (residual water content <1 wt% measured by Karl-Fischer titration) [65]. But the attapulgite binder needs approximately 500

°C for getting converted into a mechanically stable structure [53, 59]—a temperature which can already be critical for the structural stability of the zeolite [66, 95]. Due to the limited hydrothermal stability [67–69, 95] of hydrophilic zeolites, the activation step has to be carried out carefully. To avoid (partial) zeolite destruction during the activation step, the released water needs to be removed from the shape as fast as possible in order to avoid the appearance of hydrothermal conditions. The higher the temperature and higher the moisture concentration directly at the zeolite centers the higher the probability for (partial) zeolite destruction [70]. By applying a controlled temperature increase, a low height of the granules bed and/or a dry purge gas can be met these conditions. Depending on the particular conditions, a residual water content of less than one percent can be reached. During the cooling step as well as during the storage of the activated material, the presence of humidity or other adsorptive components should be avoided to prevent undesired adsorption on the material. Therefore, activated zeolite molecular sieves are usually packed in hermetically sealed packaging units, as, for example, sealed steel drums or big bags equipped with sealed inliners.

The mentioned use of a binder system opens a large variety of zeolite molecular sieves with different properties. Depending on the final application, the properties of the single shapes or the bulk material have to be optimized. For instance, with regard to the particle size, two limits should be considered: The bigger the single shape, the higher the probability for a possible limitation of the mass transfer within the shape. That means the diffusion within the single shape might be so slow that the active adsorptive matter is not completely used if the full adsorption/desorption cycle is very fast (e.g., in (vacuum) pressure swing processes [71]) [72]. But the smaller the single shapes, the larger the pressure drop over the fixed bed and the higher the risk of fluidization. Therefore, the adjustment of the size of the shapes is a critical parameter in the design of an adsorption process. Besides the size also other characteristics of the bulk material are to be adjusted such as mechanical properties such as dust formation caused by attrition/abrasion between shapes or the shapes and the wall of the adsorber vessel. Because dust may cause problems in a running adsorption process (e.g., blocking of valves), the dust formation has to be avoided by applying an optimal type and amount of binder as well as by generating single shapes with as smooth as possible surfaces. To receive smooth surfaces on single shapes of bulk material, the shaping process has to be run accurately. So, for example, in the case of extruded material an optimal plasticity of the extrusion mixture (pre-mixture), and in the case of agglomeration granulation, a post-granulation surface polishing step may help avoid/lower attrition/dust formation. Besides dusting, the mechanical stability (crush resistance) of the shapes has to be considered. The higher the bulk bed in the adsorber the bigger the weight and force that affects the lowest single shapes in the bed, which could finally result in a destruction of the shapes, and hence, generation of fines and an increased pressure drop. The binding mechanism of activated binder-containing zeolite molecular sieves is based on the generation of a network of binder material, wherein the zeolite crystals are embedded. **Figure 2** shows SEM pictures of the raw materials. The zeolite crystal agglomerates (top left) and the binder (top right) form a physically strong bound shape, which is demonstrated in **Figure 2** downright. As mentioned earlier, the use of (adsorption inert) binder material reduces the adsorption capacity zeolite molecular sieves by approximately the percentage of binder in the shape (see **Table 1**).

	Zeolite NaX powder	Binder-containing zeolite NaX beads
Water adsorption capacity at 55% relative humidity, 25°C/wt%	31.5	26.5
N_2 adsorption capacity at 1000 mbar, 25°C/cc(STP)/g	10.7	8.0
CO_2 adsorption capacity at 333 mbar, 25°C/cc(STP)/g	120.7	95.4

Table 1. Adsorption capacities for zeolite NaX (pure zeolite NaX powder and binder-containing bulk material with 82.5 wt% zeolite NaX and 17.5 wt% attapulgite).

In addition to the mechanical properties, the type and amount of binder (and the shaping process applied) have an essential influence on the structure of the secondary pore system of the shape—the part of the shape through which the transport of the molecules to and from the zeolites crystals within the shape is realized [66, 73]. Said secondary (or transport) pore system strongly influences the kinetics of the adsorbent, and hence, the adsorption process. So the zeolite NaX beads mentioned in **Table 1** show an attrition of <0.1 wt% (determined using Chemiewerk Bad Köstritz GmbH method) but relatively tight transport pores, which can lead to a slower diffusion within the zeolite beads. Using a lower amount of binder material, the adsorption capacity increases and the transport pores become slightly wider; see **Figure 3**. A faster diffusion is possible. But using a lower amount of binder material leads to an increased attrition value (0.2 wt%, determined using Chemiewerk Bad Köstritz GmbH method); for some applications that might be a high risk for an operating plant.

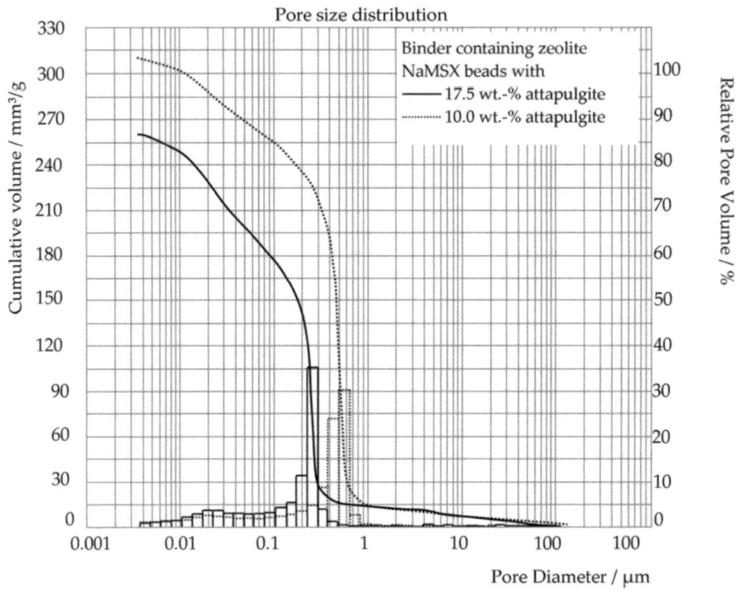

Figure 3. Hg intrusion curves of binder-containing zeolite NaX bulk material with 17.5 and 10.0 wt% of attapulgite binder.

4. Binderless zeolite molecular sieves

To avoid the above-mentioned disadvantages of binder-containing zeolite molecular sieve bulk materials such as adsorption capacity reduction by adsorption inert binders or influence of the secondary pore structure by the binder material, the so-called binderless zeolite molecular sieves were developed. There are different manufacturing procedures for binderless molecular sieves described in the open or patent literature. In most of those processes, the same shaping principles as mentioned earlier for the manufacturing of binder-containing zeolite molecular sieve are applied. They differ in the raw materials, respectively, the composition—especially the type and amount of the so-called temporary binder—of the starting pre-mixture, or the conditions for the conversion of the temporary binder into zeolite matter. Important for the generation of binderless zeolite molecular sieve is that the temporary binder contains only those elements, which are present in the target zeolite matter; that means in most of the cases silicon, aluminum, and sodium, for example, for binderless molecular sieves of A, NaX, and NaY type (or mixtures thereof). Only for the preparation of binderless zeolite LSX-type molecular sieves, potassium-containing systems are used [74].

Thus, it is possible to generate binderless zeolite shapes using temporary binder material such as kaolin as starting material [75]. A mixture of temporary binder material such as kaolin, metakaolin or silica, and zeolite powder or zeolite filter cake is also mentioned [76–83]. The conversion step can be a wet chemical [75, 79] or an at least partially autogenic thermal reaction [84].

Figure 4 shows the principle manufacturing procedure starting with shaping a pre-mixture using the above-mentioned technologies agglomeration granulation, spray granulation, or extrusion. Taggart [75] reported about shaping a mixture of kaolin and sodium hydroxide followed by drying of the shapes. It should be considered that kaolin has good binding properties but nearly no chemical reactivity. Due to that fact, the kaolin needs to be converted thermally (about 600 °C) into the chemically more reactive metakaolin. Depending on the target zeolite type, a suitable reaction solution comprising of sodium, aluminum, and/or silicon components is prepared. Usually, the shapes are aged and further processed in that reaction solution to convert the temporary binder into the desired zeolite matter. Finally, the shapes are washed, dried, and activated. Taggart mentioned that, due to the limited accessibility of the interior of the shapes by the mentioned reaction solution (blocked or too tight secondary pore system), the degree of the conversion of the temporary binder into zeolite matter, and hence, the adsorption capacity of the resulting material is limited, and can be enhanced if zeolite powder is used in the starting mixture. However, the mechanical stability of the resulting shapes is lower. Therefore, Howell et al. [85] propose the utilization of pore-forming agents (spacers), which are burned out in the following thermal treatment, thereby leaving space for the better accessibility of the interior of the shapes. Goytisolo et al. [76], Barrer et al. [77], Nozemack et al. [78], Brandt et al. [79], and Schumann et al. [80] described the manufacturing of binderless zeolite molecular sieves using a mixture of metakaolin, zeolite powder of such zeolite type, in which the metakaolin should be converted into, and additional sodium and silicon compounds such as sodium hydroxide, silicic acid or sodium silicate as feed for the shaping process. The zeolite powder in the mixture has obviously crystallization triggering properties and supports the generation of an open, for the reaction solution accessible pore system. It has to be considered that metakaolin has practically no binding properties. Thus, one has to make sure, that the process is carried out in a way, that the shapes, which are coming out of the shaping process, are mechanically stable enough until the conversion of the non-

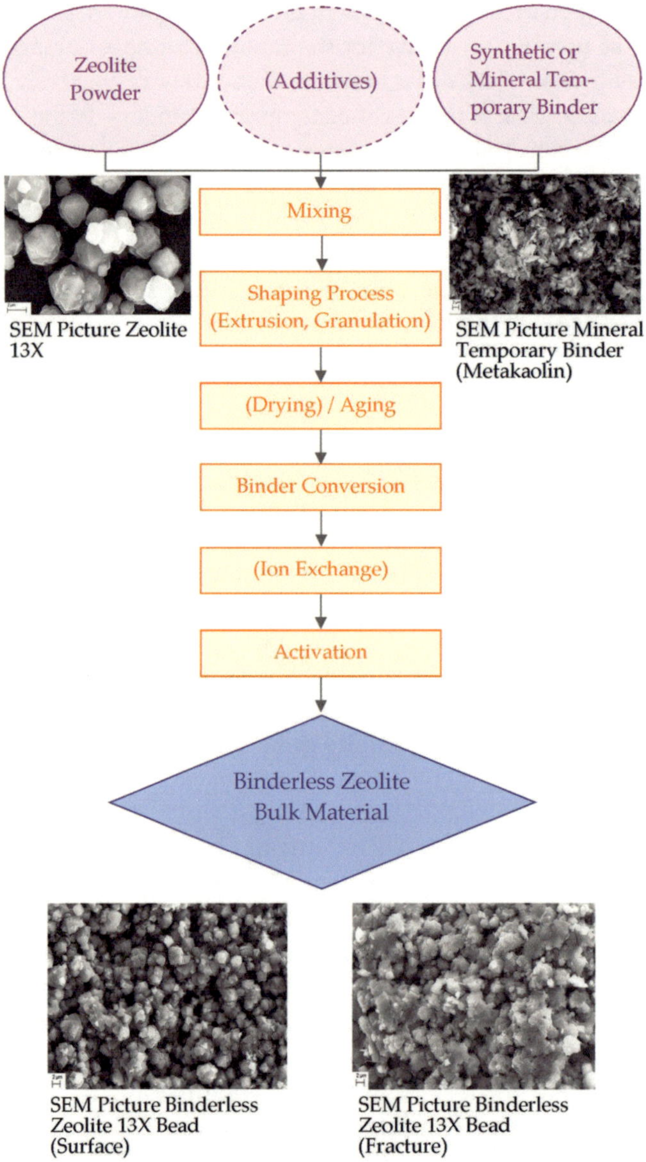

Figure 4. Scheme of the principle manufacturing procedure of binderless zeolite molecular sieve shapes (bulk material).

zeolitic compounds into zeolite matter in the reaction solution is completed. Said conversion is the basis for the mechanical stability of binderless molecular sieve shapes (see below). If it is possible to put all necessary compounds into the starting mixture for the shaping the reaction solution can be water [77, 78], otherwise all missing components for the desired zeolite formation have to be present in the reaction solution [75, 79, 80]. Another way to manufacture binderless zeolite molecular sieves is the use of silica as synthetic temporary binder. The shapes obtained are aged to achieve a certain water stability [81] followed by the conversion of the temporary binder in a solution consisting of aluminum and sodium components [82, 83].

The wet chemical reaction for the conversion of the temporary binder is preferably carried out by recirculating the reaction solution over the bulk material at suitable temperatures without moving the single shapes in order to avoid attrition between the still relatively soft shapes.

In the course of the chemical conversion of the non-zeolitic components into zeolite matter, the binding mechanism changes. In the case of kaolin/metakaolin-based systems, the mechanical stability of the shapes is now based on intergrowths between zeolite crystals in the single shape (see **Figure 4** bottom). Said intergrowths are formed during the conversion of the metakaolin into a polycrystalline zeolite matter [86]. Such unusual crystallization behavior can be explained by considering the available space for crystallization in a shape. On the outer surface of a single shape, there is enough space for a conventional and epitaxial crystal growth (see **Figure 4** down left for zeolite NaX), whereas in the interior, the space is obviously limited in a way that typical zeolite crystals such as octahedrons or cubes with rounded corners cannot be formed, but polycrystalline structures consisting of zeolite in untypical shape only (see **Figure 4** bottom right for zeolite NaX) [86]. After the chemical conversion, the bulk material is separated from the mother liquor (which can reused for further reactions) and washed until a desired pH or conductivity of the effluent is achieved. Optionally, an ion exchange (applying a suitable ion exchange solution) can be affiliated. Finally, the material is dried and thermally activated in a suitable device (e.g., rotary kiln or belt calciner). As the mechanical stability of binderless molecular sieve shapes is already achieved during the temporary binder conversion step said thermal treatment is just needed to activate the zeolite, and not anymore to set a binder component. Thus, usually lower temperatures can be applied, which saves costs and is more gentle to the thermally sensitive zeolite structure. Despite the different manufacturing procedures of binder-containing and binderless zeolite bulk material, the mechanical stability of the resulting shapes is similar. In the case products produced according to [79] and [80], a higher mechanical strength for bigger binderless zeolite spheres in comparison with binder-containing zeolite spheres of the same size and zeolite type is observed. Using comparable analytical methods for both kinds of zeolite bulk material, the attrition/dust formation is also similar.

As can be shown exemplary in **Table 2**, the static adsorption capacity of binderless zeolite NaX bulk material is similar to that of pure zeolite NaX powder. That indicates that the binderless zeolite bulk material consists of nearly 100% of zeolite matter.

	Zeolite NaX powder	Binder-containing zeolite NaX beads	Binderless zeolite NaX beads
Water adsorption capacity at 55% relative humidity, 25°C/wt%	31.5	26.5	31.2
N_2 adsorption capacity at 1000 mbar, 25°C/cc(STP)/g	10.7	8.0	10.5
CO_2 adsorption capacity at 333 mbar, 25°C/cc(STP)/g	120.7	95.4	119.4

Table 2. Adsorption capacities for zeolite NaX (pure zeolite NaX powder, binder-containing bulk material with 82.5 wt% zeolite NaX and 17.5 wt% attapulgite and binderless zeolite NaX bulk material).

A further interesting observance is, that in the above-described special manufacturing process for binderless molecular sieves, a very open secondary pore system is generated, which allows faster kinetics as compared to conventional binder-containing zeolite molecular sieves with the (usual) binder content of around 20% [86], **Figure 5** proves the presence of a larger amount of macro pores. While the ratio of meso-pore to the sum of meso- and macro-pores is 30% for conventional binder-containing zeolite molecular sieves (pre-mixture consisting only of zeolite and attapulgite) the same ratio is only 5% for binder-less zeolite molecular sieves (produced according to [80]) [66].

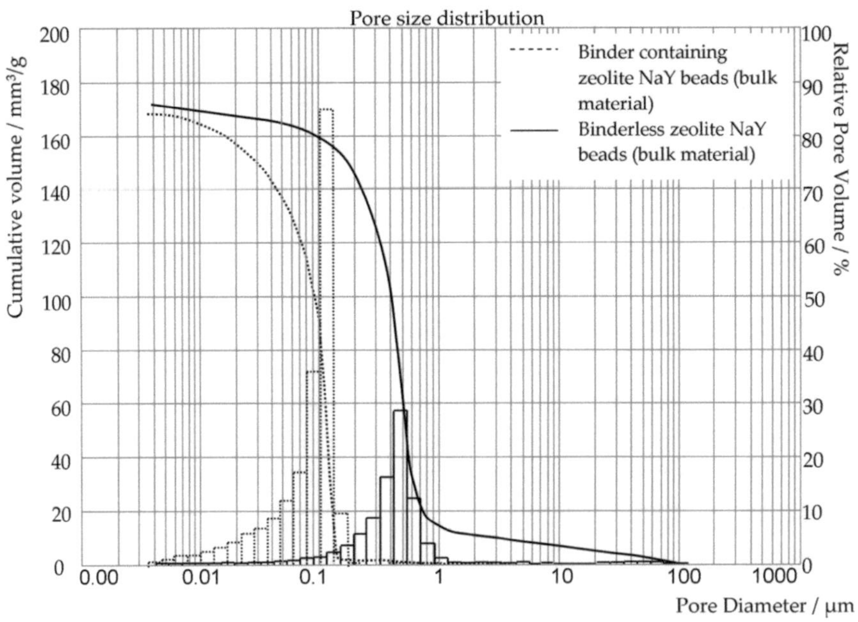

Figure 5. Hg intrusion curves of binder-containing and binderless zeolite NaY bulk material.

Due to the mentioned open secondary pore system, the above-mentioned problems concerning the inner shape transport limitations in binder-containing shapes depending on the size are eased in the case of binderless shapes.

It should, however, besides all obvious advantages of binderless over binder-containing molecular sieves mentioned earlier, noted, that, at least in the case of the kaolin/metakaolin-based systems, the structure of the binderless shapes is more rigid (less flexible) than the one of related binder-containing shapes. This can easily be explained by the different binding mechanisms: intergrowths in the case of the binderless structures, embedment in the case of binder-containing structures. This is a fact which must not be underestimated and needs to be taken into consideration especially in applications, which are connected with structural compressions/expansions (e.g., as a result of fast heat release and/or lattice changes by adsorption/desorption).

5. Compact zeolite bodies

In order to improve the overall effectivity of a given adsorption process, the increase of the volume-related adsorption capacity of the adsorbent used is often desirable. Said volume-related adsorption capacity is depending on the (bulk) density of the bulk material and on the amount of adsorption active matter within the shape (binder-containing or binderless zeolite bulk material). The bulk density itself depends on the density of the single shape(s) and on the void space between the shapes. It is known, that, for example, the packing density of a close-packing of spheres of equal diameter is ca. 74%; thus, the void space is about 26%. Theoretically, an increase of the packing density (and thus, of the bulk density) can be achieved by mixing spheres of optimal adjusted sizes, but it is certainly very difficult and expensive to produce and arrange such bed consisting of spheres of optimal sizes, the more, as the influence of the packing density on the pressure drop has always to be considered. An alternative to dense packed bulk beds can be compact zeolite bodies such as multi-channel tubes or honeycombs. By choosing appropriate geometries of such compact zeolite bodies with optimal channel dimensions a space filling (packing density) of much more, the mentioned 74% (theoretically up to 100% in the case of a void free compact body) is possible. In addition, such compact bodies can be modified in many ways to meet the requirements of the related adsorption process. Thus, a gas flow can be guided straight through the parallel arranged channels of a honeycomb-like structure, but through alternately blocking of the channels, a compulsorily perfusion is also possible [35].

Most of the manufacturing procedures of such compact zeolite bodies described in the literature are based on the use of (adsorption inert) binder material. As in the above-mentioned cases of binder-containing zeolite bulk material, synthetic binder [87] and mineral binder materials [88] are applied in the case of compact shapes as well. Consequently, the same disadvantageously impacts on the properties of the related adsorbents need to be considered: The introduction of the mentioned binder material causes a dilution of the adsorptive active matter and consequentially a reduction of adsorption capacities, and the binder material could affect the desired adsorption and desorption processes by forming a disadvantageous secondary pore system [66], an effect which at least partly can be cured by applying thermally removable pore-forming substances [89], and the fact, that impurities present in mineral binders may support undesired chemical side reactions.

At that point, it seems to be worth to remember, what was discussed above in this regard—to overcome the disadvantages of binder-containing molecular sieves bulk material. The solution (at least to a certain extent) was the application of binderless systems. And recently it could clearly be demonstrated that the same principle concept works in the case of zeolite-containing compact bodies as well: A high volume-related adsorption capacity can be reached [35]. The related honeycombs, multi-channel tubes, cylinder, etc. exhibit a remarkable adjustable high space filling (up to 100%), but at the same time a similar open secondary pore system such as mentioned earlier for binderless zeolite bulk material [90, 35].

For both types of compact zeolite bodies (binder-containing and binderless structures), the production process starts with mixing the ingredients zeolite powder, (temporary) binder, water and if needed shaping aids (such as e.g., glycerol or cellulose derivates) or in case of binder-containing compact bodies pore-forming substances if necessary (see **Figure 1**). Shaping technologies can be extrusion, pressing, or molding—depending on the desired geometry of the compact zeolite body. It should be noted that in the case of the binderless systems, the application of kaolin instead of metakaolin as main non-zeolitic component has been shown as being advantageous. In case of the binder-containing compact zeolite bodies, the following and final manufacturing step is the thermal treatment for drying and activation, where the organic compounds are burned out as well as the all water and other adsorbed compounds present in the system is removed and the binder is set. In case of binderless compact zeolite bodies, a first thermal treatment is needed for removing the organic shaping aids and converting kaolin into metakaolin [35, 91]. The generated metakaolin is converted into zeolite during a following wet chemical reaction. The related reaction solution consists of all ingredients necessary for the conversion of the metakaolin into the desired zeolite type (A, X or NaY) similar to the manufacturing of binderless zeolite bulk material. The conversion can be followed by further modification processes, for example, ion exchange. The completeness of the conversion of the non-zeolitic into zeolite matter can be demonstrated, for example, by adsorption measurements (in combination with XRD [35]). The static water adsorption capacity for zeolite 4A honeycombs in comparison with the fully crystalline zeolite 4A powder is shown in **Table 3** proving that the honeycomb consists of nearly 100% zeolite 4A matter.

	Zeolite 4A powder	**Binderless zeolite 4A honeycomb**
Water adsorption capacity at 55% relative humidity, 25°C/wt%	24.8	24.7

Table 3. Adsorption capacities for zeolite 4A (pure zeolite 4A powder and binderless zeolite 4A honeycomb).

Just for completion it should be noted that compact zeolite-containing bodies can also be manufactured by washcoating [92, 93], where mainly for catalytic applications, the catalytically active zeolite component is deposited onto monolithic support structures, as, for example, ceramic (in most cases cordierite) honeycombs by washcoating. The result is a compact body with a thin zeolite layer. Furthermore, the crystallization of zeolite matter directly on

http://dx.doi.org/10.5772/63442

aluminum foams is published [94]. Such structures exhibit interesting flow characteristics, but due to the rather large void fraction and a relatively thin zeolite layer onto the aluminum support, a low volume-related zeolite amount, and hence, a low volume-related adsorption capacity is observed.

6. Conclusion

Zeolite molecular sieves play an important role as highly effective adsorbents in major industrial adsorption processes. The properties of such molecular sieves can be tailored according to the requirements of the related applications. Besides the common binder-containing molecular sieves, binderless grades are applied. Due to the particular binding mechanism and the related special secondary pore system, binderless molecular sieves are mainly applied where fast kinetics is required, thus, especially in (vacuum) pressure swing adsorption processes. Furthermore, the principle concept for the manufacturing of binderless zeolite bulk material can be applied to the manufacturing of compact zeolite bodies, too. This opens ways for a further improvement of the related adsorption processes.

Acknowledgements

The authors gratefully acknowledge Gundula Fischer, Hannes Richter, and Andreas Häusler from Fraunhofer IKTS Hermsdorf, Germany for the contribution to this work.

Author details

Kristin Gleichmann*, Baldur Unger and Alfons Brandt

*Address all correspondence to: k.gleichmann@cwk-bk.de

Chemiewerk Bad Köstritz GmbH, Bad Köstritz, Germany

References

[1] Grand View Research, Inc.: Global Zeolite Molecular Sieve Market Analysis, Market Size, Regional Outlook, Application Analysis, Competitive Insights and Forecasts, 2015 To 2022, 2015

[2] QY Research Zeolite Molecular Sieve Research Center: Global and Chinese Zeolite Molecular Sieve Industry Report, 2015

[3] The Catalyst Group: The Industrial Adsorbents Business, 2013

[4] K. Shah, Journal of the Combustion Society of Japan, 2013, 55, 171, 13–20

[5] M. Last, A. Finn, UK Shale Gas Processing: Part One, Engineering Tomorrow...Today, Costain group PLC, 2015

[6] M. Last, A. Finn, UK Shale Gas Processing: Part Two, Engineering Tomorrow...Today, Costain group PLC, 2015

[7] J. Tobin, P. Shambaugh, E. Mastrangelo, Natural Gas Processing: The Crucial Link Between Natural Gas Production and Its Transportation to Market, Energy Information Administration, Office of Oil and Gas, January 2006

[8] Gas Technology: Technical Information about Natural Gas Cleaning and Treatment, Gas- und Umwelttechnik GmbH, Pörner Gruppe

[9] S. Mokhatab, P. Meyer, Selecting Best Technology Lineup for Designing Gas Processing Units, Gas Processors Association – Europe, Europe Sour Gas Processing Conference, Sitges, Spain 13th–15th May 2009

[10] S. Nicolas, P.-G. Schmitt, US Patent 2002 / 0 141 940, 2002

[11] „...wir packen Feuchtigkeit", Technische Information, Tropack Packmittel GmbH, Lahnau-Waldgirmes, without year

[12] DIN 55417, Packhilfsmittel – Trockenmittelbeutel – Anwendung, 1997

[13] B. Müller, W. Rath, Formulierung von Kleb- und Dichtstoffen, Hannover: Vincentz Network, 2004, 218 pp.

[14] R. Miller, M. R. Miller, Air Conditioning and Refrigeration, New York: The McGraw-Hill Company, Inc., 2006, 148 pp.

[15] T. Tsuchiya, S. Ide, T. Shibanuma, EP Patent 0 974 633, 2000

[16] H. Meyer, R. Doclo, J. Seynaeve, DE Patent 2 823 211, 1979

[17] D. Bathen, M. Breitbach, Adsorptionstechnik, 1st ed., Berlin, Heidelberg: Springer, 2001

[18] K. Haruna, K. Ueda, M. Inoue, H. Someda, EP Patent 0 334 495, 1989

[19] D. A. Lagree, F. W. Leavitt, US Patent 4 810 265, 1989

[20] B. Su, M. Bulow, J. Bein, A. F. Ojo, S. Jale, D. Shen, Q. M. Wang, F. R. Fitch, US Patent 6 350 298, 2002

[21] A. J. Kidnay, W. R. Parrish, Natural Gas Processing, CRC Press, Taylor & Francis Group, Boca Raton 2006

[22] K. Schuman, B. Unger, A. Brandt, Chemie Ingenieur Technik, 2010, 82, 6, 929–940. doi: 10.1002/cite.201000013

[23] M. S. Peters, K. D. Timmerhaus, R. E. West, Plant Design and Economics for Chemical Engineers, 5th ed., New York: McGraw-Hill, 2003, 109 pp.

[24] D. Broughton, C. Gerhold, US Patent 2 985 589, 1961

[25] S. Sohn, S. Kulprathipanja, J. Rekoske, US Patent 6 407 305, 2002

[26] M. Minceva, A. E. Rodrigues, in 2nd Mercosur Congress on Chemical Engineering, 4th Mercosur Congress on Process Systems Engineering, Rio de Janeiro 2005

[27] T. Mori, T. Tsuchiya, Y. Noguchi, T. Yamamoto, I. Takeuchi, K. Matsumura, S. Akita, K. Sando, M. Kanaoka, S. Otani, T. Iwamura, US Patent 3 761 533, 1973

[28] C. Wetter, E. Brügging, F. Doetkotte, Machbarkeitsstudie zur Absolutierung von Bioethanol aus landwirtschaftlichen Brennereien, Münster 2005

[29] A. Aden, M. Ruth, K. Ibsen, J. Jechura, K. Neeves, J. Sheehan, B. Wallace, L. Montague, A. Slayton, J. Lukas, National Renewable Energy Laboratory, Technical Report, Golden, CO 2002, 38 p.

[30] W. Schulz, Wärmespeicher für die Energieeffizienz, Energiespeichersymposium, Stuttgart, Germany, 7 March 2012

[31] P. Feddeck, F. Meyer, Heizen mit Zeolith-Heizgerät. BINE-Informationsdienst, Fachinformationszentrum Karlsruhe, 2005. ISSN 0937-8367

[32] T. Nonnen, S. Beckert, K. Gleichmann, A. Brandt, B. Unger, H. Kerskes, B. Mette, S. Bonk, T. Badenhop, F. Salg, R. Gläser, Chemie Ingenieur Technik, 2016, 88, 3, 363–37. doi:10.1002/cite.201500136

[33] M. Fauth, H. Jerg, K. Paintner, A. Reiter, R. Rieger, DE Patent 10 2008 043 550, 2008

[34] A. Jakway, H. Disch, W. Neumaier, EP Patent 2 617 343, 2013

[35] K. Schumann, B. Unger, A. Brandt, G. Fischer, H. Richter, J. Jänchen, Chemie Ingenieur Technik, 2014, 86, 1–2, 106–111, doi:10.1002/cite.201300026

[36] F. Wolf, H. Fürtig, E. Lemnitz, GDR Patent 43 221, 1965

[37] R. M. Milton, US Patent 2 882 244, 1953

[38] V. R. Balse, M. Bülow, P. Connolly, F. Fitch, A. F. Ojo, US Patent 5 932 509, 1999

[39] J. G. Moscoso, US Patent 9 133 037, 2015

[40] D. Täschner, A. Pritzl, DE Patent 10 2009 012 532, 2010

[41] R. M. Milton, US Patent 2 996 358, 1961

[42] R. J. Ross, J. D. Sherman, US Patent 4 666 693, 1987

[43] H. Tschritter, K. Voigtberger, B. Gojdar, D. Gruhle, B. Unger, DE Patent 10 2006 002 092, 2007

[44] O. Grubner, P. Jírů, M. Rálek, Molekularsiebe, Berlin: VEB Deutscher Verlag der Wissenschaften, 1968, 42 p.

[45] L. Qiang, Z. Ying, C. Zhijun, G. Wei, C. Lishan, Petroleum Science, 2010, 7, 403–409. doi:10.1007/s12182-010-0085-x

[46] A. V. Tekhtamysheva, L. D. Konoval'chikov, Chemistry and Technology of Fuels and Oils, 1990, 26, 397–399

[47] M. L. Pavlov, O. S. Travkina, R. A. Basimova, I. N. Pavlova, B. I. Kutepov, Petroleum Chemistry, 2009, 49, 36–41

[48] J. Jänchen, T. Herzog, K. Schumann, A. Brandt, B. Unger, poster: Properties of a novel binderless Y-type zeolite for thermal adsorption storage application, 17th International Zeolite Conference, Moscow, Russia, 06–11 July 2013

[49] V. H. Grassian, S. C. Larsen, Dekker Encyclopedia of Nanoscience and Nanotechnology. In: J. A. Schwarz, C. I. Contescu, K. Putyera (eds.), 2nd ed., New York: CRC Press, 2004, pp. 1137–1146

[50] S. Sang, Z. Liu, P. Tian, Z. Liu, L. Qu, Y. Zhang, Materials Letters, 2006, 60, 1131–1133. doi:10.1016/j.matlet.2005.10.110

[51] D. W. Breck, Zeolite Molecular Sieves Structure, Chemistry, and Use, New York: John Wiley & Sons, 1974, 339 p.

[52] D. R. Corbin, A. J. Sacco, S. L. Suib, Q. Zhang, US Patent 7 014 837, 2006

[53] F. Ullmann, W. Foerst, Encyklopädie der technischen Chemie, Volume 2, Urban & Schwarzenberg, 1970, 511 p.

[54] W. Foerst, Ullmanns Encyklopädie der technischen Chemie, Volume 17, Wiley, 1979, 14 p.

[55] K. C. Weston, P. Purcell, US Patent 8 680 344, 2014

[56] A. K. Ghosh, C. Mihut, M. Simmons, US Patent 9 180 441, 2015

[57] J. G. Heinrich, C. M. Gomes, Einführung in die Technologie der Keramik – Vorlesungsmanuskript TU Clausthal, Available from: [Accessed: 2016-03-04]

[58] W. J. Mitchell, W. F. Moore, US Patent 2 973 327, 1961

[59] D. Jaussaud, K. Weston, A. Pfenninger, B. Kleeb, US Patent 6 743 745, 2004

[60] O. Grubner, P. Jírů, M. Rálek, Molekularsiebe, Berlin: VEB Deutscher Verlag der Wissenschaften, 1968, 43–44 pp.

[61] Y. Chang, S. N. Waughn, P. R. Stafford, E. Nebesh, B. Slawski, WO Patent 2005/056184, 2005

[62] E. Bartholomé et al., Ullmanns Encyklopädie der technischen Chemie, Volume 17, Wiley, New York, 1979, 14 p.

[63] O. Grubner, P. Jírů, M. Rálek, Molekularsiebe, Berlin: VEB Deutscher Verlag der Wissenschaften, 1968, 44 p.

[64] L. R. M. Martens, M. J. Janssen, N. Coute, J. R. Lattner, US Patent 7 754 933, 2010

[65] O. Grubner, P. Jírů, M. Rálek, Molekularsiebe, Berlin: VEB Deutscher Verlag der Wissenschaften, 1968, 46 p.

[66] K. Schumann, A. Brandt, B. Unger, F. Scheffler, talk: Adsorptionseigenschaften bindemittelfreier Zeolith X-Formkörper, Jahrestreffen des Fachausschusses Adsorption, Würzburg, Germany, 24–25 March 2011

[67] D. W. Breck, Zeolite Molecular Sieves Structure, Chemistry, and Use, New York: John Wiley & Sons, 1974, 492–493 pp.

[68] J. Jänchen, A. Brandt, J. Schmeißer, B. Unger, H. Stach, U. Hellwig, poster: Water sorption studies on novel binderless molecular sieves X and A for heat storage and heat transformation, 16th International Zeolite Conference, Sorrento, Italy, 4–9 July 2010

[69] P. Meyer, paper: Hydrothermal damaging of molecular sieve and how to prevent it, Gas Processors Association Europe, Paris, France, Feb. 2003

[70] D. W. Breck, Zeolite Molecular Sieves Structure, Chemistry, and Use, New York: John Wiley & Sons, 1974, 490–495 pp.

[71] C. Benkmann, US Patent, 4 326 858, 1982

[72] F. Bretschneider, C. Van Lookeren, M. Nebelung, H. Klemm, WO Patent 03/040259, 2003

[73] D. Bathen, M. Breitbach, Adsorptionstechnik, 1st ed., Berlin: Springer, 2001, 3–5 pp.

[74] H. Shigeru, A. Harada, EP Patent 1 142 622, 2001

[75] R. L. Taggart, G. L. Ribaud, US Patent 3 119 659, 1964

[76] J. A. Goytisolo, D. D. Chi, H. Lee, US Patent 3 906 076, 1975

[77] R. M. Barrer, P. J. Denny, Journal of the Chemical Society Hydrothermal chemistry of the silicates. Part IX. Nitrogenous aluminosilicates, 1961, 971–982. Doi: 10.1039/JR9610000971

[78] R. J. Nozemack, C. W. Chi, J. J. Schwonke, US Patent 4 381 255, 1983

[79] A. Brandt, J. Schmeißer, B. Unger, H. Tschritter, U. Henkel. B. Gojdar, D. Gruhle, G. Winterstein, DE Patent 10 2008 046 155, 2009

[80] K. Schumann, A. Brandt, B. Unger, DE Patent 10 2012 010 109, 2012

[81] L. Puppe, G. Ulisch, DE Patent 3 401 485, 1985

[82] G. Heinze, DE Patent 1 203 238, 1965

[83] B. Hees, L. Puppe, G. Reiss, US Patent 5 962 358, 1999

[84] D. E. Hildebrandt, US Patent 4 381 256, 1983

[85] P. A. Howell, N. A. Acara, US Patent 3 119 660, 1964

[86] K. Schumann, B. Unger, A. Brandt, F. Scheffler, Microporous and Mesoporous Materials, 2012, 154, 119–123. doi:10.1016/j.micromeso.2011.07.015

[87] H.-G. Fritz, C. Trefzger, H.H. Höfer, DE Patent 198 15 564, 1999

[88] T. Takeuchi, M. Mouri, S. Okabayashi, S. Miyamura, US Patent 5 387 564, 1995

[89] E. Sextl, R. Eckehart, P. Kleinschmit, A. Kiss, DE Patent 42 02 671, 1992

[90] J. Jänchen, T. H. Herzog, K. Gleichmann, B. Unger, A. Brandt, G. Fischer, H. Richter, Performance of an open thermal adsorption storage system with Linde type A zeolites: Beads versus honeycombs, Microporous and Mesoporous Materials, 2015, 207, 179–184. doi:10.1016/j.micromeso.2015.01.018

[91] K. Schumann, B. Unger, A. Brandt, G. Fischer, H. Richter, DE Patent 10 2012 020 217, 2012

[92] H. Keshavan, O. P. Siclovan, D. Hancu, US Patent 2009/0318283, 2009

[93] Y. Huang, A. Bar-Ilan, US Patent 6 759 358, 2004

[94] F. Scheffler, R. Herrmann, W. Schwieger, M. Scheffler, Preparation and properties of an electrically heatable aluminium foam/zeolite composite, Microporous and Mesoporous Materials, 2004, 67, 53–59. doi:10.1016/j.micromeso.2003.10.006

[95] K. Schumann, A. Brandt, B. Unger, F. Scheffler, Chemie Ingenieur Technik 2011, 83, No. 12, 2237–2243. doi:10.1002/cite.201100151

Chapter 6

Hollow Titanium Silicalite Zeolite: From Fundamental Research to Commercial Application in Environmental-Friendly Catalytic Oxidation Processes

Changjiu Xia, Min Lin, Bin Zhu, Yibin Luo and Xingtian Shu

Additional information is available at the end of the chapter

http://dx.doi.org/10.5772/63699

Abstract

The systematical investigation on the synthesis, characterization, formation mechanism, and catalytic application of hollow titanium silicalite (HTS) zeolite has been reviewed. HTS is prepared through a "dissolution–recrystallization" post-treatment in the presence of template under hydrothermal conditions. Compared with TS-1, HTS is of unique hollow voids and with high framework Ti content, which significantly increase the mass diffusion and the amount of active sites, respectively. Thus, HTS zeolite displays high catalytic activity and stability in many oxidation processes with H_2O_2 oxidant, that is, cyclohexanone ammoximation, phenol hydroxylation, propylene epoxidation, Baeyer-Villiger oxidation of cyclohexanone, and selective oxidation of aromatics and cycloalkanes. The former three ones have been commercialized and run smoothly, which have promising economic and environmental significance.

Keywords: Hollow Titanium Silicate, Environmental friendly, Post synthesis, Selective oxidation, Commercialization

1. Introduction

Zeolites are regular microporous materials composed by TO_4 tetrahedra (T stands for Al, Si, Sn, Ti, and B) through sharing the interconnected oxygen atoms in the framework positions. They possess abundant of ordered channels and cages within molecular dimensional range, which can reversibly uptake the molecules (smaller than 1 nm) depending on the structure topology

[1–3]. Aluminosilicate zeolites display Brönsted acidity, due to the isomorphous substitution of tetrahedral Si atom by trivalent Al atoms in the framework positions, needing one proton ion to keep electron charge balance. As a result, Si–Al zeolites are widely used as acid catalysts, adsorbents, and ion-exchanges in petroleum refining and chemical manufacture [4–6]. Furthermore, to extend the application scope of zeolite, transition metal ions have been incorporated into the framework sites, with the endowment of redox property [7–9]. Up to now, the MFI-typed titanium silicalite-1 (TS-1) zeolite is the most successful and profoundly investigated one in the area of isomorphously substituted heteroatom zeolites. Since it is originally claimed by Enichem in 1983, TS-1 zeolite has attracted intensive attentions from both academic and industrial practices, owing to its high catalytic performance in hydrocarbon oxidation process. That is because the framework Ti species can enhance the nucleophilicity capability of H_2O_2 molecules to attack electrophilic groups during heterogeneous catalysis, through the acceptor–donator interaction between the empty 3*d* orbits of tetrahedral Ti species and electron pairs in H_2O_2 molecules [10–13]. Compared with traditional oxidation processes, TS-1 catalyzed routes are much more environmental friendly, with H_2O as the only side product. For example, the one-step cyclohexanone ammoximation approach by direct reacting with H_2O_2 and NH_3 has already replaced the conventionally four-step hydroxylamine routes in commercial ε-caprolactam production units [14–17].

However, at the early stage, TS-1 zeolite usually exhibited poor catalytic activity and reproducibility, due to the mismatch of hydrolysis and crystallization rates between Si and Ti species under hydrothermal conditions. To solve this problem, hollow titanium silicate (HTS) zeolite was developed by Lin and his coworkers at SINOPEC in 1990s, through a post-synthesis treatment in the presence of organic structure directing agents (OSDAs) [18–21]. It is found that HTS zeolite is of unique hollow cavities inside crystal, which are preferentially to enhance the mass diffusion property of reagents and products during catalytic reactions. Besides, the post-treatment promotes the hydrated condensation between Ti-OH and Si-OH groups, forming much more framework Ti species. Thus, HTS zeolite shows much higher catalytic performance than TS-1 zeolite in many oxidation reactions, such as phenol hydroxylation, cyclohexanone ammoximation, and propylene epoxidation and aromatic oxidation. On the other hand, HTS zeolite has its own intellectual property, which is very important to overcome patent restriction for its commercial production. Since 2000, SINOPEC has successfully completed the commercialization of HTS zeolite at a scale of 100 t/a. After that, HTS zeolite was initially used as catalyst for the environmental friendly cyclohexanone ammoximation process in the 140 kt/a commercial manufacture of ε-caprolactam at Baling Branch, SINOPEC, to replace the four-step HPO process, with low formation of toxic waste and side-products [22–25]. Thus, this route has been considered as one of the most important cases in green chemistry, due to its high atom utilization efficiency and low source consumption. Moreover, the HTS zeolite has also been employed in other industrial oxidation reactions, such as phenol hydroxylation and propylene epoxidation, with low concentration H_2O_2 solution as oxidant. Very recently, the 100 kt/a industrial demonstration plant of propylene epoxidation has been run smoothly, with high propylene oxide yield and H_2O_2 conversion.

http://dx.doi.org/10.5772/63699

Above all, HTS zeolite shows interesting physicochemical properties and great performance in the catalytic oxidation process, and we have done abundant of fundamental investigations to understand the basic principles in this topic, such as the hollow cavity characterization, chemical status of Ti species and the formation mechanism of constricted hierarchical structure. However, until now, the synthesis, characterization and catalytic application of HTS zeolite is still not reviewed systematically, especially for its catalytic performance in industry. Herein, we try to describe the main achievements on all aspects of HTS zeolite in brief, and we do hope it can provide a novel view for modifying zeolite materials to meet specific needs.

2. Synthesis and characterization of HTS zeolite

2.1. Synthesis of HTS zeolite through the "dissolution-recrystallization" treatment

TS-1 zeolite is the first crystalline transition metal inserted microporous materials (also referred to as heteroatom zeolites). As reported by Taramasso et al., TS-1 zeolite was synthesized using tetraethyl orthosilicate (TEOS), tetrapropylammonium hydroxide (TPAOH), and tetrabutyl orthotitanate (TBOT) as raw materials, under hydrothermal conditions and autogeneous pressure [18, 20]. The high pH value of TPAOH favors the existence of isolated Ti species coordinated by low aggregated Si species. The most important recipe for preparing particularly active TS-1 zeolite lies in the accurate control of the hydrolysis of Ti and Si sources in aqueous alkaline solution, to promote the formation of Si–O–Ti bonds, rather than Ti–O–Ti bonds. Although many alternative synthesis routes had been proposed by researchers over the past 30 years, except for Enichem Company, it is still difficult to prepare the active and stable TS-1 zeolite through direct method [26–29]. Consequently, to improve the catalytic performance of TS-1 zeolite, a novel post-synthesis treatment method has been designed by the researchers at SINOPEC in 1990s. In a typical procedure, the parent TS-1 zeolite is synthesized following the classic Enichem route as previously mentioned; and then the as-made TS-1 zeolite is calcined at high temperature in air to remove the organic templates inside micropores. In the next step, the calcined TS-1 zeolite is washed by specific additive agents, for trying to remove the extraframework Ti species (majorly in the form of anatase TiO_2); after that, this TS-1 zeolite is mixed with aqueous TPAOH solution, and sealed in a Teflon autoclave. Then, the autoclave is continuously heated at high temperature (usually over 120°C) under statistic state for several days, with the formation of white HTS zeolite powder. It is worthy noted that the as-made TS-1 zeolite, which containing TPAOH template, cannot be used as precursor to prepare HTS zeolite treated under the same procedure as calcined TS-1 zeolite. Thus, it is inferred that part of TS-1 zeolite is dissolved in the alkaline solution, and the soluble Si and Ti species are recrystallized under the impact of TPAOH molecules during the post-synthesis processing. Most importantly, this post-synthesis treatment is very convenient and available in both laboratory and industry, on which allows the commercial manufacture of HTS zeolite at large scale.

2.2. Confirmation of hollow cavities and reincorporation of Ti species

It is well known that the intracrystalline porosity is of great importance to enhance the catalytic performance of zeolites during heterogeneous reactions [30–32]. Different from TS-1 zeolite, HTS zeolite shows a typical H_2 hysteresis loop at the region of P/P_0 between 0.45 and 1 in the N_2 physisorption curve (**Figure 1**), which suggests the existence of ink-bottle alike hierarchical pores. It is attributed to the tensile strength effect, owing to the difference between capillary condensation for adsorption and capillary evaporation for desorption. The hysteresis loop starting from $P/P_0 = 0.45$ usually reveals the constricted mesopores and/or macropores inside zeolite crystal [33–35]. Thus, to further confirm the feature of hierarchical pores, transmission electron microscopy (TEM) method has been used, as illustrated in **Figure 2**. Obviously, abundant of hollow cavities (around several to tens nm in size) have been observed in the HTS zeolite, while the conventional TS-1 zeolite is completely solid. Moreover, to learn the three dimensional distribution of hollow cavities, HTS zeolite was characterized by using electron tomography (ET) technology [36–38]. As displayed in **Figure 2b** (insert), intracrystalline cavities are highly dispersed inside HTS zeolite with different size and shape. Consequently, we can infer that intracrystalline pores have been created via the post-synthesis treatment in the presence of template under hydrothermal conditions. Herein, it is noted that HTS zeolite is still of high crystallinity and uniform morphology, which are favor of the high stability during catalytic process. Therefore, this post-synthesis route takes much more advantages than base/acid leaching method and provides a novel strategy to design and modify zeolite materials for meeting specific demands in heterogeneous catalysis.

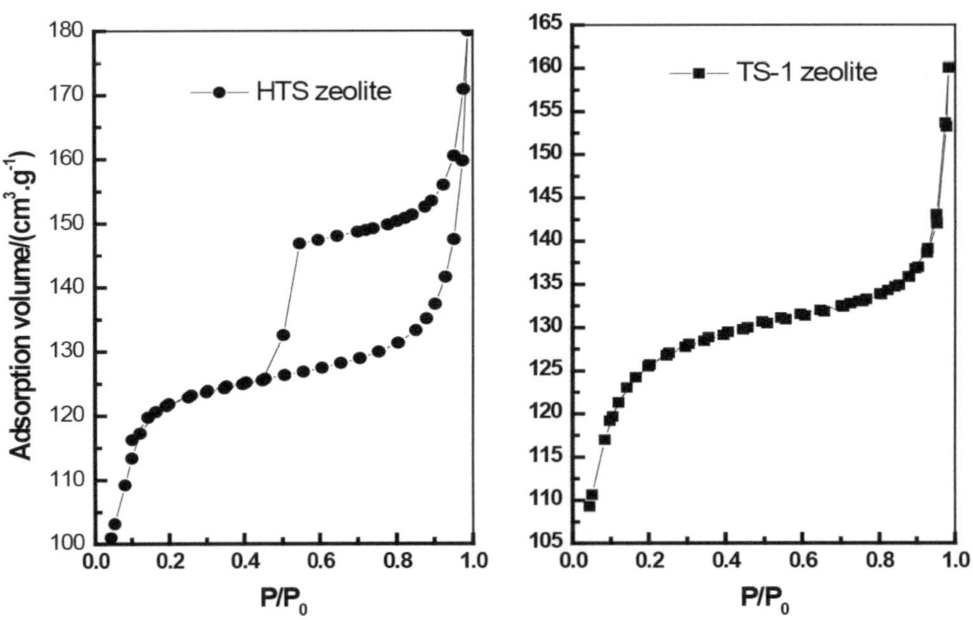

Figure 1. Low-temperature N_2 adsorption-desorption curves of HTS and TS-1 zeolites.

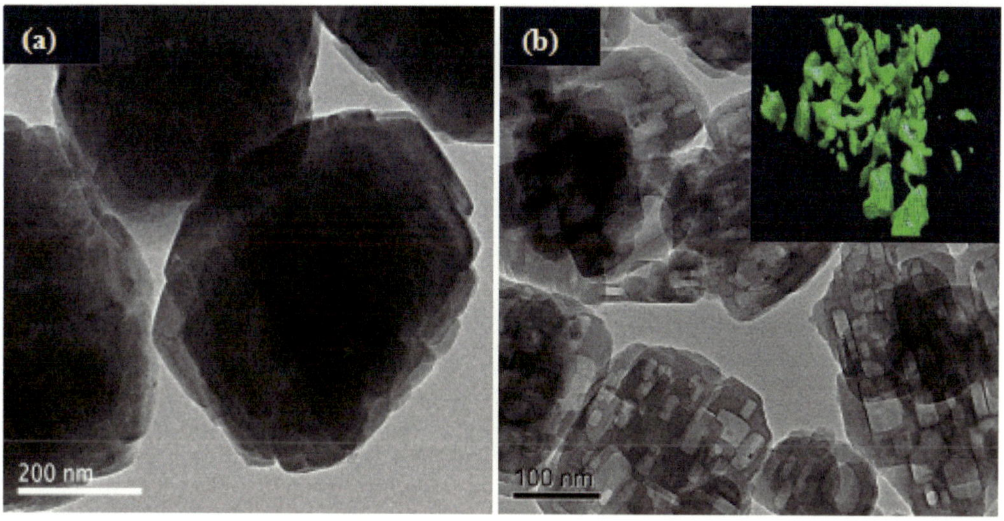

Figure 2. (a) TEM image of conventional TS-1 zeolite, (b) TEM image of HTS zeolite, and the restructured image of hollow cavities (insert).

Besides the formation of hollow cavities, the reincorporation of Ti species into framework positions has also been found in our previous work. Firstly, the chemical status of Ti species in the HTS zeolite has been distinguished by X-ray photoelectron spectroscopy (XPS) through peak fitting treatment [39, 40]. Apparently, two kinds of Ti species, referred to as framework Ti and extraframework Ti species, have been detected in the external surface of HTS zeolite (as presented in **Figure 3**). That is because the chemical environmental of Ti atoms in the framework positions are different from that in the extraframework Ti species, especially for octahedral anatase TiO_2 crystal.

Additionally, it is widely accepted that the band at around 960 cm^{-1} in the FT-IR spectrum is attributed to the stretching vibration of Si–O–Ti bonds, indicating the incorporation of tetrahedral framework Ti species. And the band intensity ratio between 960 and 800 cm^{-1} (I_{960}/I_{800}) is relevant with the framework Ti content [41, 42]. Thus, it is concluded more tetrahedral Ti species have been inserted into framework sites after the post-synthesis treatment, due to the higher I_{960}/I_{800} value of HTS than that of TS-1 zeolite (as shown in **Figure 4**). Furthermore, to verify the reincorporation of Ti species, ^{29}Si MAS NMR spectroscopy has been employed to detect the chemical environmental of Si species, which can indirectly give the information on the chemical status of Ti species. Generally, the bands at around −113 and −116 ppm are ascribed to the framework $Si(OSi)_4$ and $Si(OTi)(OSi)_3$ Species, respectively, owing to induction effect of framework Ti atoms substituted Si atoms [43–45]. As shown in **Figure 5**, HTS zeolite has a larger peak at −116 ppm than TS-1 zeolite, and thus it is well demonstrated that "dissolution-recrystallization" treatment favors the reincorporation of tetrahedral Ti species into framework sites.

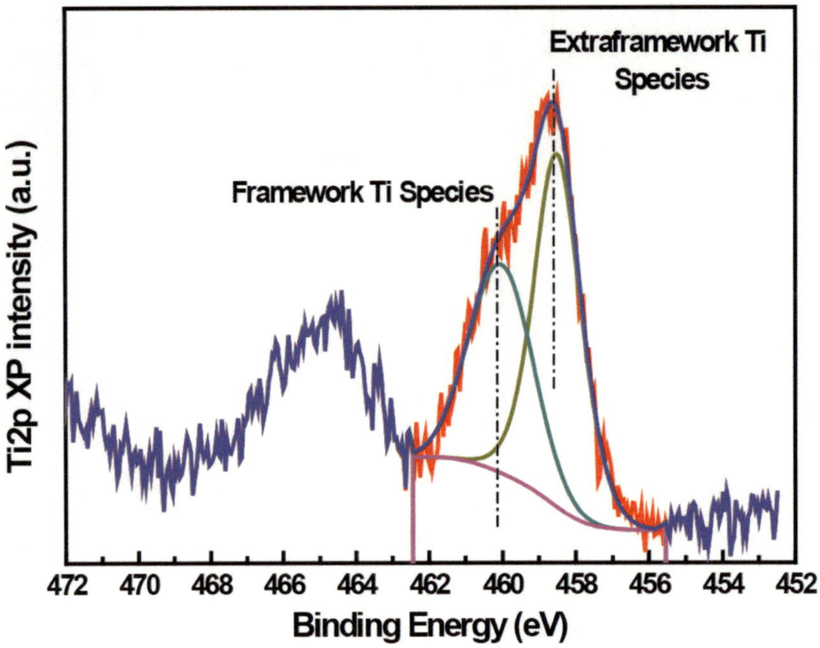

Figure 3. XPS spectrum of HTS zeolite with peak fitting to confirm framework and extraframework Ti species.

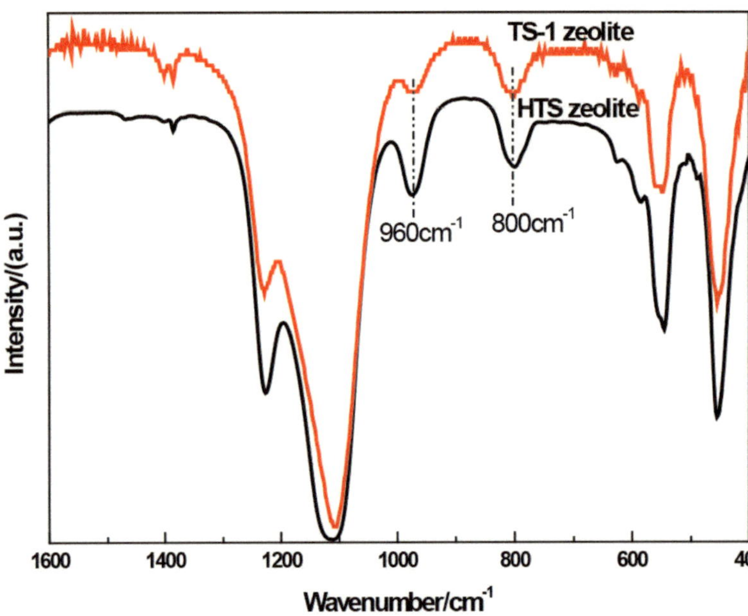

Figure 4. FT-IR spectra of both TS-1 and HTS zeolites.

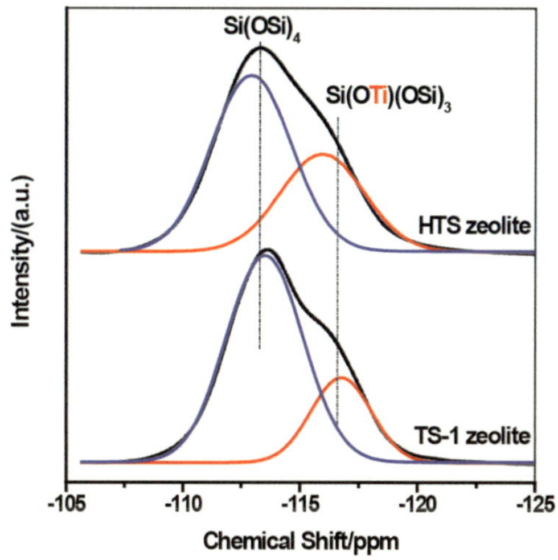

Figure 5. ^{29}Si MAS NMR spectra of both TS-1 and HTS zeolites.

2.3. The formation mechanism of hollow voids inside HTS zeolite

The detailed formation mechanism of hollow cavities inside zeolite is still not fully understood, but huge investigations on this post-synthesis treatment have been carried out. To certify the "dissolution-recrystallization" process, the as-made (organic template containing) TS-1 zeolite was treated by using three different methods, as following: (a) the as-made TS-1 zeolite was post-synthesized in the TPAOH solution at high temperature; (b) the as-made TS-1 zeolite was calcined to remove organic template, and then treated in the TPAOH solution at 80°C in air;

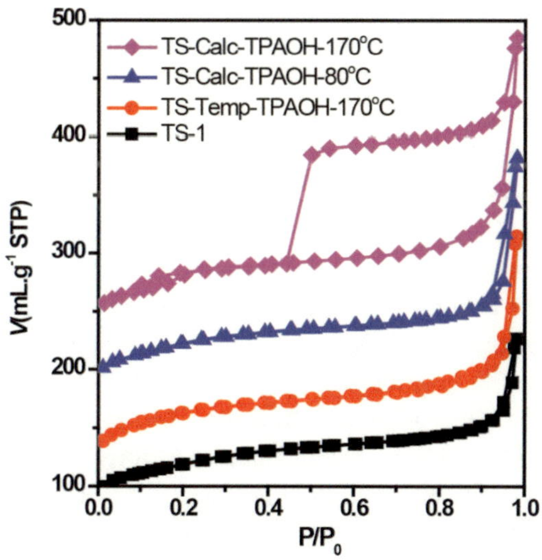

Figure 6. Low-temperature N_2 adsorption–desorption curves of different titanium containing zeolites.

(c) the as-made TS-1 zeolite was calcined to remove organic template, and then post-synthesized in the TPAOH solution at 170°C under autogeneous pressure. **Figures 6** and **7** show the N_2 physisorption curve and TEM characterization results of these mentioned samples. We can observe that only the calcined TS-1 zeolite can be used as starting material to prepare HTS zeolite under hydrothermal conditions.

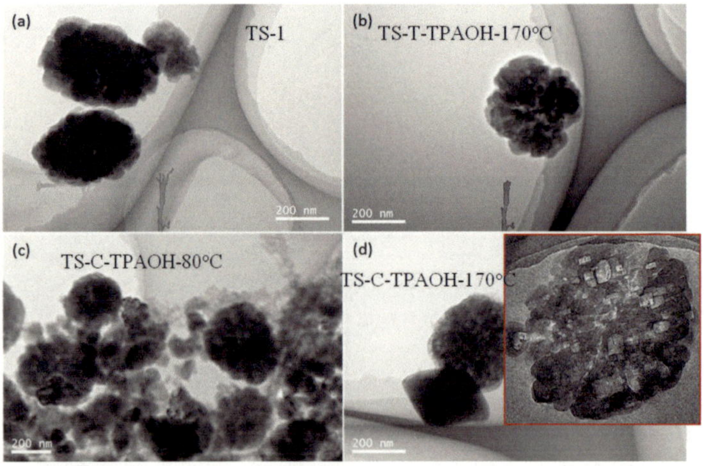

Figure 7. TEM images of different titanium containing zeolites.

On the formation mechanism of hollow cavities, Mielby et al. has ever used cetyltrimethylammonium bromide (CTAB) alkaline solution as organic structure directing agent (OSDA) to produce the Silicalite-1 zeolite with hollow voids [46–48]. They consider that the CTAB molecules can enter into the bulk body of zeolite to form micelles, which severs as the mesoporous template for hollow structure via directing the soluble Si and Ti species dissolved from zeolite crystal. However, this viewpoint is not fully reasonable enough to demonstrate the generation of hollow cavities inside HTS zeolite, owing to the larger molecular diameter of TPAOH than that of CTAB molecules. On the other hand, CTAB is of poor capability to cause the formation of MFI topology. Based on these mechanistic models, one plausible mechanism has been proposed to explain the formation of hollow cavities inside zeolite during "dissolution–recrystallization" process, but it still need further detailed analysis results to support it. We consider that, in the initial stage, the calcined TS-1 zeolite can be dissolved in the TPAOH solution to, owing the hydrolysis of T–O–T bonds (T stands for the framework atoms) catalyzed by OH^- anions in both external surface and bulk body of zeolite. It is worthy to note that only OH^- anions, rather than TPA^+ ions, go into zeolite body along microporous channels. Due to the absence of TPA^+ ions, the recrystallization of soluble Ti and Si sources cannot be occurred, without the formation of new TS-1 inside zeolite crystal. At the same time, the "dissolution–recrystallization" process happens in the external surface, with the changing of zeolite morphology. That is why the hollow voids cannot be produced for the post-synthesis of template-containing TS-1 zeolite, which is ascribed to the restriction

http://dx.doi.org/10.5772/63699

of mass transfer of OH^- group within zeolite. Above all, this proposed mechanism provides a novel view for describing the formation hollow structure, which is in favor of the inner diffusion of raw materials and products during catalytic oxidation processing [49–51].

3. Applications of HTS zeolite in catalytic oxidation processes

3.1. Cyclohexanone ammoximation process

Cyclohexanone oxime, produced by cyclohexanone ammoximation, is one of the most important chemical intermediates to produce ε-caprolactam via a Beckmann intramolecular rearrangement [52–54]. ε-caprolactam is widely used to manufacture nylon-6 worldwide, which acts as the precursor for producing fiber, plastic, resins, so on. As the fast development of human society, especially for recent 10 years, the demands of ε-caprolactam and cyclohexanone oxime are becoming rapidly increasing. The traditional cyclohexanone oxime production methods majorly focus on the hydroxylamine phosphate (HPO) or hydroxylamine sulfate (HSO) routes, as shown in **Figure 8a** [55–57]. Obviously, it is very complex, highly polluted and corrosive to prepare hydroxylamine, which is used as oxidant for cyclohexanone ammoximation reaction, due to the usage of strong acid solutions and ammonia. Large amount of low value-added side-products (particularly ammonium sulfate) are formed in this approach, and thus the traditional methods obey the fundamental principles of green chemistry and need to be remarkably improved in both academia and industry.

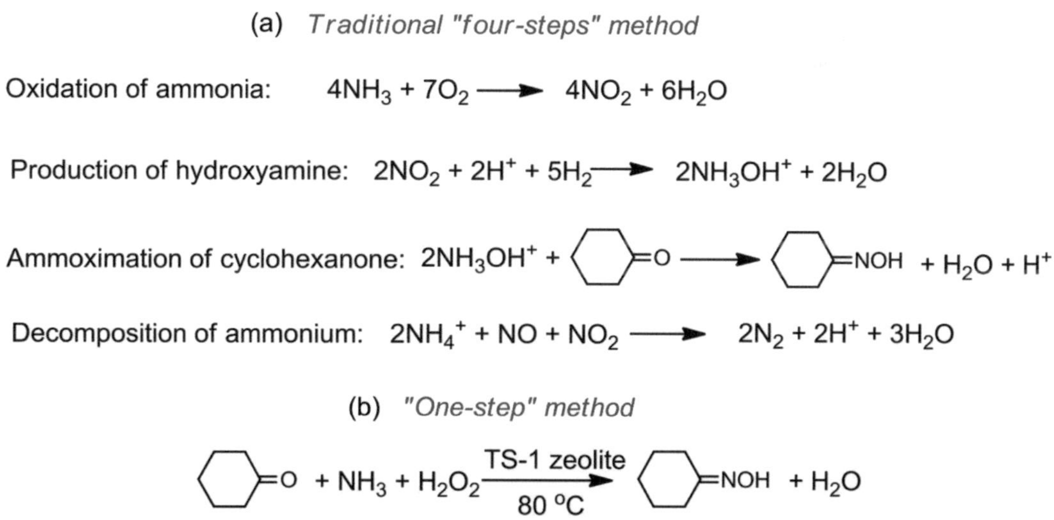

Figure 8. Reaction pathways of traditional HPO route (a) and one-step route catalyzed by TS-1 (b).

To overcome these drawbacks in HPO and HSO routes, a new liquid phase one-step process was developed by Enichem in 1987 [58–60]. It is based on the "one-step" cyclohexanone ammoximation of catalyzed by TS-1 zeolite, with ammonia and hydrogen peroxide solutions as reactants, as demonstrated in **Figure 8b**. After that, extensive researches have been carried out with make this novel route become economic and technological available in industry. Compared with traditional routes, there is no hazard side-product generated in the TS-1 catalyzed route, with water as the only by product and the atom efficiency over 85%. Here, the cyclohexanone oxime can be achieved by one-step method, which is operated under mild condition, with low reaction temperature and low pressure. Therefore, cyclohexanone ammoximation reaction catalyzed by TS-1 zeolite is considered as a completely environmental-friendly process.

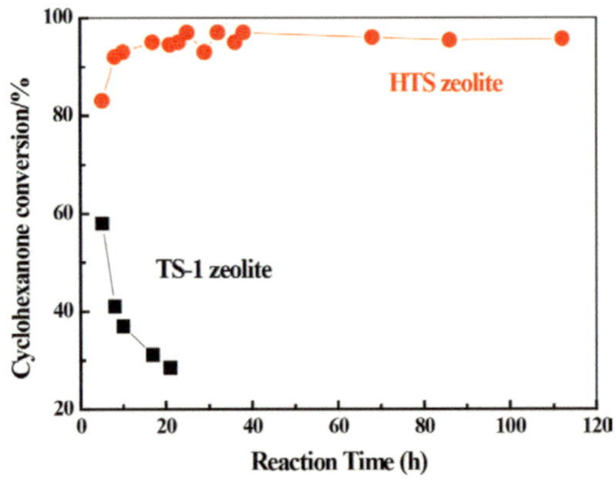

Figure 9. Catalytic performance of TS-1 and HTS zeolites in the cyclohexanone ammoximation with low concentration H_2O_2 solution as oxidant.

However, the TS-1 zeolite synthesized by traditional method is usually of poor catalytic performance in cyclohexanone ammoximation process, due to the presence of alkaline ammonia, causing the hydrolysis of Ti–O–Si bonds. Thus, in the early years, the stability of TS-1 zeolite was considered as the main obstacle to prevent this process be commercialized at large scale, which was attributed to the low lifetime of TS-1 zeolite in ammoximation system. That is why only first one12 kt/a cyclohexanone ammoximation demonstration installation was established by Enichem in 1994 at Porto Margherita, Italy [61]. With respect to the deactivation of TS-1 zeolite, three main aspects were proposed by Petrini et al., such as (a) pore filling by heavy organic side-products, (b) the transformation of framework Ti atoms into extraframework Ti species, and (c) dissolution of framework Si atoms in the alkaline solution [62–65]. And very recently, we have confirmed this phenomenon using multiple characterization methods and catalytic evaluation. Therefore, to enhance the stability of TS-1 zeolite in ammoximation, the HTS zeolite was developed by Lin et al., at SINOPEC through a novel post-synthesis of conventional TS-1 zeolite in the presence of TPAOH under hydrothermal condition. The catalytic performance of HTS and TS-1 zeolites in cyclohexanone ammoximation process is outlined in **Figure 9**, which suggests that HTS zeolite displays much higher catalytic activity and longer lifetime than conventional TS-1 zeolite under the same reaction conditions. In our opinion, the fast cyclohexanone conversion rate is attributed to the generation of hollow voids inside crystal, which favors the accessibility of raw material molecules to tetrahedral framework Ti sites, and improves inner mass diffusion, thus reducing the formation of heavy organic deposit. And the high stability of HTS zeolite is associated with the reincorporation of Ti species into the framework positions via the dehydrated condensation between Ti-OH and Si-OH groups under hydrothermal conditions. As shown in **Figure 9**, the lifetime of HTS zeolite in ammoximation reaction is much longer than that of TS-1 zeolite, with high cyclohexanone conversion (>95%) for over 110 h. As a result, the commercial cyclohexanone ammoximation units catalyzed by HTS zeolite had been builded in Baling Petrochemical Company, SINOPEC at 140 kt/a scale since 2000, to replace the conventional four-step HPO process. In industrial process, the cyclohexanone conversion and oxime selectivity are all higher than 99.5%, whereas the utilization efficiency of H_2O_2 is about 90%. It is worthy to note that the size of HTS zeolite is about 200–350 nm, which means these particles are hard to be separated from reaction system. Thus, the separation process of HTS zeolite by inorganic member was been successfully developed and widely applied in industrial ammoximation units. In a word, this novel route takes promising advantages for the manufacture of ε-caprolactam, with significant economic and environmental meaning. Additionally, in very recent years, several cyclohexanone ammoximation units catalyzed by HTS zeolite have been established or are still under construction, with even larger scale (>200 kt/a) in China.

3.2. Phenol hydroxylation process

Phenol hydroxylation to prepare catechol (*o*-dihydroxybenzene) and hydroquinone (*p*-dihydroxybenzene), which are widely applied to manufacture agrichemical, photography chemicals, antioxidants, polymerization inhibitors, is one of most important reactions in fine chemical industry [66, 67]. Traditionally, catechol, usually together with hydroquinone, is produced by oxidation of phenol with various oxidants, such as 60–70% aqueous H_2O_2 solutions, performic acid, and ketone peroxides (such as α-hydro hydroperoxides). Heavy transition metal compounds and strong phosphoric acid are selected as catalysts for these hydroxylation reactions, but they are very difficult to be separated from products [68, 69]. Obviously, these processes are very complicated and far beyond environmental friendly to be carried out at commercial scale, with very low atom utilization efficiency. There are huge amount of liquid and/or solid phase wastes need to be disposed, which are highly toxic and corrosive for the installations. Fortunately, one completely green and clean phenol hydroxylation process via using TS-1 zeolite as catalyst has been developed by EniChem, which uses low concentration (usually 30 wt%) H_2O_2 solution, and is carried out under benign conditions (lower than 100°C and atmosphere pressure). **Figure 10** shows the comparison of catalytic performance of HTS and TS-1 zeolite in phenol hydroxylation under the same conditions. We observe that HTS zeolite can faster catalyze the transformation of phenol than TS-1 zeolite, which is correlated with its inner hierarchical structure. That is to say, hollow cavities promote the mass diffusion inside zeolite crystal, while the higher phenol conversion is dependent on the more framework Ti content, as previously characterized. Based on the excellent catalytic performance, several phenol hydroxylation installations have been builded and run smoothly in China.

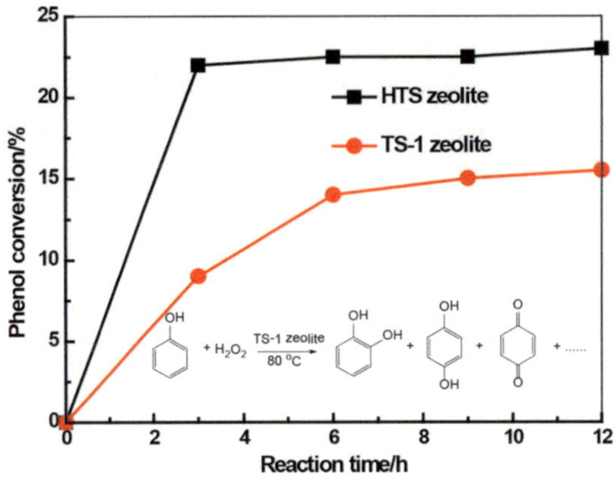

Figure 10. Catalytic performance of HTS and TS-1 zeolite in phenol hydroxylation reaction.

Furthermore, the phenol hydroxylation is chosen as probe reaction to reveal the kinetic activity of framework Ti sites, owing to its mechanism on the activation of H_2O_2, rather than functional groups in phenol. Although high-concentration (>60 wt%) H_2O_2 solutions are of strong nucleophilic attack capability during catalytic oxidation reaction, it is considerably difficult and dangerous to store, transfer and use them at large scale. Interestingly, the activated H_2O_2 molecules (in the form of Ti-OOH species) by HTS zeolite are of as strong nucleophilicity as 60 wt% H_2O_2 solution or peroxide acids. And the solid HTS zeolite crystal is easy to be separated from liquid phase, and thus the heterogeneous hydroxylation process has attracted kind attentions from the researchers in both academic and industrial sections.

3.3. Propylene epoxidation (HPPO) process

Propylene oxide (PO) is also one of the most important chemical intermediates to produce polyesters and polyurethane in industry, with an increasingly huge market worldwide (over 7.5 million tons in 2010) [70–73]. Generally, two basic strategies are majorly employed for commercial PO production, including hydrochlorination route and co-oxidation route. The former one involves the transformation of propylene to chloropropanols and dehydrochlorination, which need to use the Cl_2 and $Ca(OH)_2$ as basic materials, respectively, with the formation of corrosive and low value-added wastes. The other one lies in the co-oxidation of propylene and other organic compounds (such as ethylbenzene or isobutane) with air or O_2 as oxidant, in the presence of catalyst. The coproducts in this route are t-butyl alcohol, MTBE or styrene, but the main drawback is the high capital and material cost and long reaction pathway. To avoid these disadvantages, an ultra environmental-friendly propylene epoxidation process with hydrogen peroxide as oxidant (referred to as HPPO route) was exploited by using TS-1 zeolite as catalyst, and in 2009 a new HPPO plant was started up by BASF and Dow chemical at 300 kt/a scale in Antwerp [74–77]. Later on, several HPPO plants were established in many countries, such as Korea, Thailand, and China. At the same time, the HPPO process catalyzed by HTS zeolite has been developed by SINOPEC, and one 100 kt/a propylene epoxidation industrial initialization has been established and run smoothly in China.

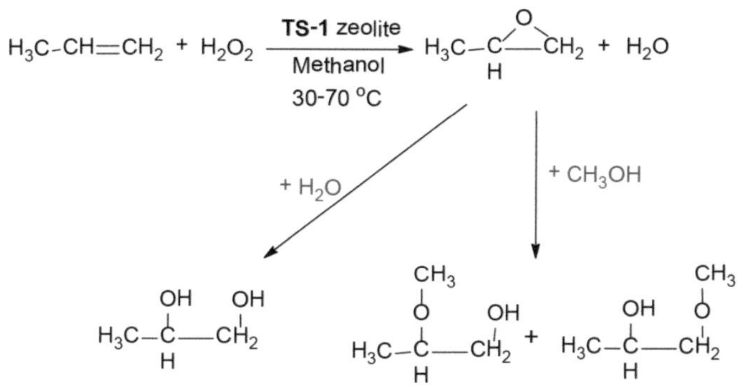

Figure 11. Reaction network of propylene epoxidation via HPPO process.

With respect to the reaction network of HPPO process, except for the major propylene epoxidation, other side reactions (as shown in **Figure 11**) are also happened, causing the decrease of PO yield. Thus, it is necessary to use large amount of methanol as solvent and operate this route at low temperature. A 1.0 kt/a HPPO process, schematic flow diagram is demonstrated in **Figure 12**. The epoxidation reaction is carried out in a fixed-bed reactor, with the mixture of propylene and H_2O_2 injecting into the reactor. When the epoxidation reaction at 1.0 kt/a scale is carried out in a wide range, T = 30–70°C, P = 0.5–2.0 MPa, WHSV of H_2O_2 molecules is 0.12–1.2 h^{-1}, $n(CH_3OH)/n(H_2O_2)$ is 5–25, n(Propylene)/$n(H_2O_2)$ is 1.2–2.5, the TOF of H_2O_2 is around 31–32 mmol g^{-1} h^{-1}(X_{H2O2} is 96–99%) and selectivity of PO is 96–99%, which is much better than that reported by BASF and DOW (the selectivity of PO is 93.2%). Furthermore, the raw propylene oxide produced in the fixed-bed reactor is purified in the PO purification column, obtaining the final PO product at 99.97% purity. And then, the purified PO has been used as chemical intermediates to manufacture 1,2-ethyl glycerol, oilfield demulsifier, propylene carbonate, triisopropanolamine, hydroxypropy, and dimethyl carbonate [78].

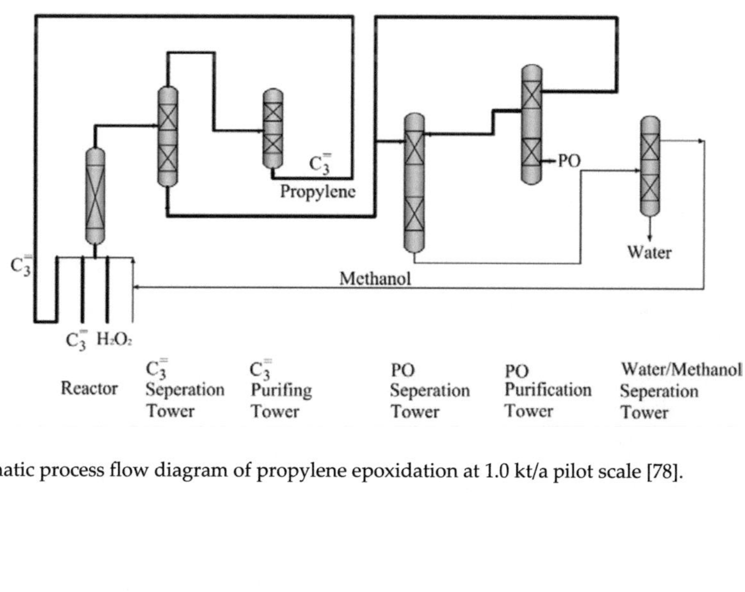

Figure 12. Schematic process flow diagram of propylene epoxidation at 1.0 kt/a pilot scale [78].

3.4. Baeyer–Villiger oxidation of cyclohexanone

Baeyer–Villiger (B–V) oxidation is an important reaction to produce esters or lactones from ketones via inserting one O atom in C–C bond. Especially, ε-carprolactone, prepared via cyclohexanone BV oxidation, is widely used to synthesize polycaprolactone and poliglecaprone [79–83]. Traditionally, many kinds of peracids, such as *m*-chloroperbenzoic acid, trifluoroperacetic acid and perbenzoic acid, are chosen as oxidizing agents and catalyst to motivate BV reaction. The main advantages are the high ketone conversion and fast transformation rate, but there are still many significant drawbacks, such as low selectivity of target products, low atomic utilization, being harmful to the environment and dangerous to handle. Thus, up to now, only three companies can commercially produce ε-carprolactone based on peracids oxidant in the world, owing its long and complex reaction pathway. To beat these drawbacks, low concentration (usually 30–50 wt%) H_2O_2 solution (which cheap and easy to handle) is considered as the best candidate to hazardous replace peracids. But it is of poor nucleophilic capability to attack the carbonyl group in ketones during BV oxidation reaction. As a result, two basic strategies are proposed to accelerate the heterogeneous B–V oxidation of cyclohexanone using aqueous H_2O_2 solution: (a) activation the carbonyl groups to make the C atoms in carbonyl group easy to be attacked by H_2O_2 molecules; (b) increasing the nucleophilic capability of H_2O_2 molecules by forming M-OOH species (M represents transition metal element, which could accept the lone electron pairs by employing their empty *d* orbits) [84, 85].

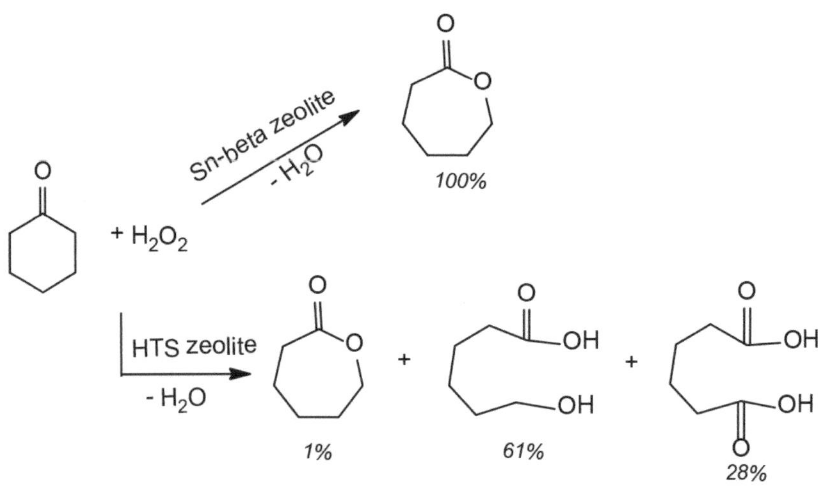

Figure 13. BV oxidation of ketones with different zeolite catalysts under mild reaction conditions [86, 87].

For activation of carbonyl group, Sn-beta zeolite reported by Corma and his coworkers [88–90] shows the most promising catalytic performance in BV oxidation of cyclohexanone with low concentration H_2O_2 solution, with almost 100% ε-carprolactone selectivity. It is attributed to the tetrahedral coordinated Sn atoms that were incorporated into the framework of beta zeolite can accept the electron pairs in carbonyl group of cyclohexanone. On the other hand, TS-1 zeolite is extensively confirmed as the Lewis catalyst to activate H_2O_2 molecules.

Consequently, HTS zeolite is taken as catalyst to promote the BV oxidation of cyclohexanone under mild conditions, as show in **Figure 13**. It suggests that HTS zeolite has the different product distribution from the Sn-beta zeolite, with high selectivity of deep-oxidative products, which are 6-hydroxyhexanoic acid and adipic acid. 6-hydroxyhexanoic acid is produced through the ring-opening reaction of lactone, due to the poor stability of 7-member ring of ε-carprolactone and high water content in reactor. Then, the C-OH groups in 6-hydroxyhexanoic acid can be further oxidized by Ti-OOH species, to obtain the C-OOH group in adipic acid [86]. Compared with Sn-beta catalyzed route, HTS-catalyzed reaction system does not use any specific organic solvents, which lower the concentration of oxidant and are high-energy consumed to be separated. Thus, it is in favor of the formation of a series of important chemical intermediates in this "one-pot" reaction. However, there are still many kinds of undesirable by-products formed at the same time, reducing the yields of target product. As a consequence, the BV oxidation of cyclohexanone catalyzed by HTS zeolite is still under investigation in the laboratory to minimize other side reactions.

3.5. Catalytically selective oxidation of cyclohexane

Cyclohexane oxidation is an ultra important reaction to product KA oil (including cyclohexanone and cyclohexanol) in industry, for manufacturing various of chemical precursors and final products, as shown in **Figure 14**. Generally, the KA oil is usually obtained via these methods: hydrogenation of phenol, hydration of cyclohexene, hydrogenation of benzene, and air oxidation of cyclohexane (>90% for KA oil production) [91–94]. The cyclohexane oxidation with air or O_2 can be carried out in the presence of various catalysts or in the absence of catalyst, with high temperature (>100°C) and autogeneous pressure. Although air is very cheap and easy to be obtained, this route also suffers serious restrictions in industrial practice, such as low yield of target product, high pollution and waste generation, and particularly high-energy consumption needed for separation of products.

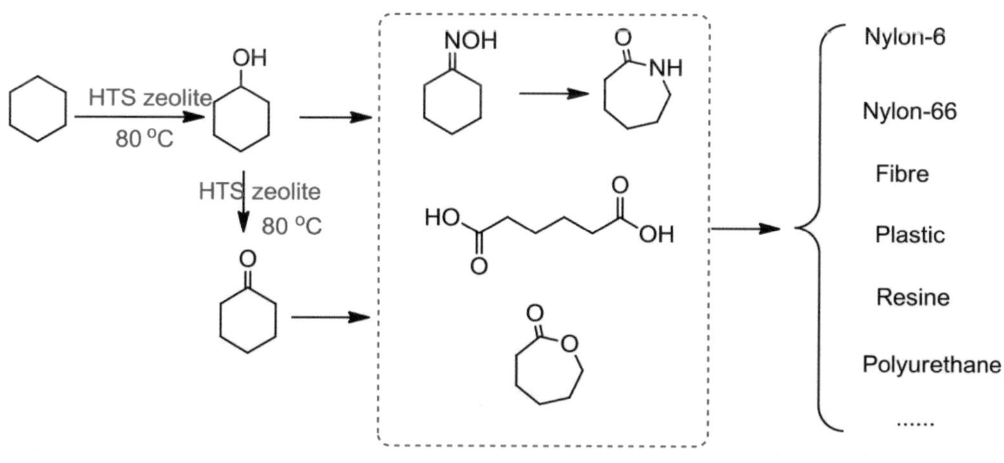

Figure 14. Reaction pathway of oxidation of cyclohexane catalyzed by HTS zeolite with H_2O_2 solution.

For solving these limitations, cyclohexane oxidation with H_2O_2 catalyzed by TS-1 zeolite has also been investigated. It is found that the KA oil can be achieved through this approach, with very high cyclohexane conversion and selectivity of KA oil under mild conditions. Herein, HTS zeolite is also introduced to catalyze the cyclohexane oxidation, as illustrated in **Figure 15**. We observe that the total yield of KA oil is increasing as a function of reaction time, which is up to 70% after reaction time more than 10 h. The optimized reaction parameters are as follow: T = 60–80°C, catalyst content = 5–10%, cyclohexane to H_2O_2 molar ratio = 0.5, acetone to cyclohexane molar ratio = 10–15, reaction time = 6–10 h. Under this conditions, the yield of KA oil is over 40%, with a cylohexanone/cyclohexanol ratio larger than 2. Above all, the cyclohexane oxidation catalyzed by HTS zeolite shows great industrial potential, and the further study on scale up of this route is still under carrying out [95, 96].

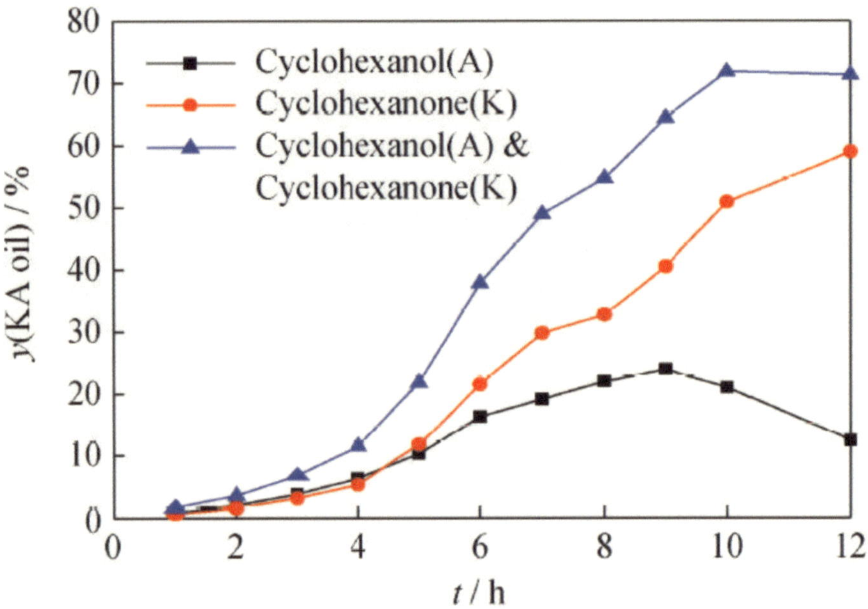

Figure 15. Catalytic performance of HTS zeolite in cyclohexane oxidation with H_2O_2 solution [95].

3.6. Catalytic oxidation of aromatic compounds

Aromatic oxidation is of great important, but still a big challenge in the chemical industry, because the kinds and quantity of synthetic aromatic compounds demanded are so huge in the world. Several synthesis methods have been developed for the oxyfunctionalization of aromatics, such as the partial oxidation of isopropylbenzene, neutralization of benzenesulfonate, hydrolysis of chlorobenzene, direct oxidation with N_2O, and direct oxidation with H_2O_2 solution [97–100]. among them, the last two routes are considered as typical green chemistry process, but the origination of N_2O is so limited.

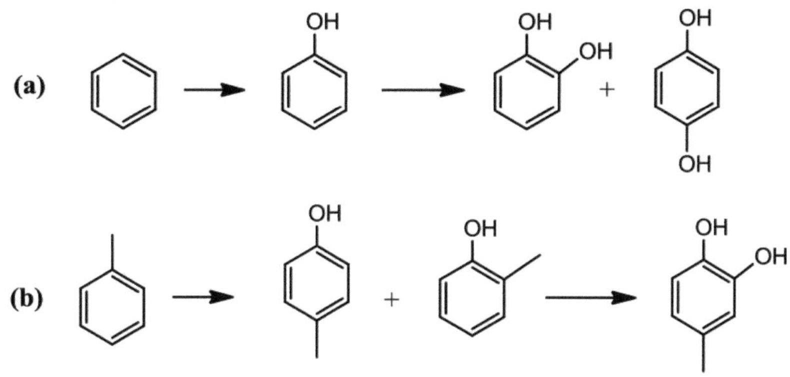

Figure 16. Reaction pathway of aromatic oxidation with H_2O_2 solution catalyzed by HTS zeolite.

In our laboratory, both benzene and toluene oxidation reactions catalyzed by HTS zeolite have been examined (**Figure 16**), which show some aromatic conversion and high selectivity of target product [101, 102]. And it is found that polar solvents are useful for high conversion rate, but the high intersolubility between aromatics and H_2O_2 is still very difficult. Thus, it is necessary to intensify the mixing inside various reactor.

4. Conclusion and perspective

In summary, HTS zeolite, which synthesized via "dissolution–recrystallization" treatment, is an innovative catalytic material for the conversion of oxygenated hydrocarbons under mild conditions. It has been well demonstrated that highly dispersed intracrystalline mesopores and/or macropores are formed inside zeolite, which favors the inner mass diffusion and high conversion rate confirmed by physicochemical adsorption of N_2 and catalytic evaluation. The mechanism of the formation of these hollow cavities has been proposed to further understand the "dissolution–recrystallization" process, which involves the diffusion of OH^- group inside zeolite crystal to form hollow voids, and the recrystallization of soluble Ti and Si sources cause the change of zeolite morphology. At the same time, the reincorporation of Ti species into framework positions has been verified using multiple characterization methods, with the significantly decreasing of framework defects (Q^3 Si species) and increasing of framework Ti content. That is why HTS zeolite shows even longer lifetime than conventional TS-1 zeolite in alkaline cyclohexanone ammoximation reaction, owing to relative prefect framework atoms tolerated to basic corrosion, except for the fast mass transfer via hierarchical structure. Since 2000, HTS zeolite has been successfully commercialized by SINOPEC in China.

Based on the unique hollow feature and high framework content, HTS zeolite displays very excellent catalytic performance and stability in many selective oxidation reactions of organic compounds, that is, cyclohexanone ammoximation, phenol hydroxylation, propylene epoxidation, Baeyer–Villiger oxidation of cyclohexanone, selective oxidation of cyclohexane and benzene, so on. The first three items have been already put into commercialization to produce bulk chemicals by SINOPEC, of great social and environmental significance to reduce pollution in industry. The others are still under investigation in the laboratory, but some ones have present good potential to replace these hazard and low atom efficiency process. For further development of HTS zeolite material and its catalytic application, we consider that the synthesis and modification of HTS zeolite should be paid more attention, because catalytic materials are the fundamentally original power for the development of heterogeneous catalysis. Moreover, the fast development of chemistry science and human society provides so many opportunities for catalytic materials with specific characteristics. Thus, from the viewpoint of application, HTS zeolite still has considerable applications in the conversion of oxygenated hydrocarbons, such as in fine chemical, environmental catalysis, and biomass utilization. Additionally, combining the redox property of HTS with other functions (such as hydrogenation of noble metal, Pt or Pd) is also a good way for updating its application. In a word, this chapter is of both theoretical and practical meanings to design and produce catalytic materials for facing specific demands. We have described a systematical investigation on zeolite synthesis and environmental-friendly heterogeneous catalysis processes from laboratory to large commercial scale.

Author details

Changjiu Xia, Min Lin*, Bin Zhu, Yibin Luo and Xingtian Shu

*Address all correspondence to: minlin@tom.com

State Key Laboratory of Catalytic Materials and Reaction Engineering, Research Institute of Petroleum Processing, SINOPEC, Beijing, People's Republic of China

References

[1] Davis M E, Lobo R F. Zeolite and molecular sieve synthesis[J]. *Chem. Mater.*, 1992, 4(4): 756–768.

[2] Freyhardt C C, Tsapatsis M, Lobo R F, et al. A high-silica zeolite with a 14-tetrahedral-atom pore opening[J]. *Nature*, 1996, 381(6580): 295–298.

[3] Davis T M, Drews T O, Ramanan H, et al. Mechanistic principles of nanoparticle evolution to zeolite crystals[J]. *Nat. Mater.*, 2006, 5(5): 400–408.

[4] Primo A, Garcia H. Zeolites as catalysts in oil refining[J]. *Chem. Soc. Rev.*, 2014, 43(22): 7548–7561.

[5] Jacobs P A, Dusselier M, Sels B F. Will Zeolite - Based Catalysis be as Relevant in Future Biorefineries as in Crude Oil Refineries?[J]. *Angew. Chem. Int. Ed.*, 2014, 53(33): 8621–8626.

[6] García-Martínez J, Li K, Krishnaiah G. A mesostructured Y zeolite as a superior FCC catalyst–from lab to refinery[J]. *Chem. Comm.*, 2012, 48(97): 11841–11843.

[7] van Bokhoven J A, Lamberti C. Structure of aluminum, iron, and other heteroatoms in zeolites by X-ray absorption spectroscopy[J]. *Coordination Chem. Rev.*, 2014, 277: 275–290.

[8] Ouyang X, Hwang S J, Xie D, et al. Heteroatom-Substituted Delaminated Zeolites as Solid Lewis Acid Catalysts[J]. *ACS Catal.*, 2015, 5(5): 3108–3119.

[9] Wang J, Xie J, Zhou Y, et al. New facile way to isomorphously substituted Cr-β zeolite and its catalytic performance[J]. *MicroporousMesoporous Mater.*, 2013, 171: 87–93.

[10] Taramasso M, Perego G, Notari B. Preparation of porous crystalline synthetic material comprised of silicon and titanium oxides[P]. US 4410501, 1983.

[11] Bonino F, Damin A, Ricchiardi G, et al. Ti-peroxo species in the TS-1/H_2O_2/H_2O system[J]. *J. Phys. Chem. B*, 2004, 108: 3573.

[12] Munakata H, Oumi Y, Miyamoto A. A DFT Study on Peroxo-Complex in Titanosilicate Catalyst: Hydrogen Peroxide Activation on Titanosilicalite-1 Catalyst and Reaction Mechanisms for Catalytic Olefin Epoxidation and for Hydroxylamine Formation from Ammonia[J]. *J. Phys. Chem. B*, 2001, 105: 3493.

[13] Bordiga S, Damin A, Bonino F, et al. The Structure of the Peroxo Species in the TS-1 Catalyst as Investigated by Resonant Raman Spectroscopy[J]. *Angew. Chem. Int. Ed.*, 2002, 114: 4928.

[14] Corma A, García H. Lewis acids as catalysts in oxidation reactions: from homogeneous to heterogeneous systems[J]. *Chem. Rev.*, 2002, 102: 3837.

[15] Dartt C B, Davis M E. Catalysis for environmentally benign processing[J]. *Ind. Eng. Chem. Res.*, 1994, 33: 2887.

[16] R. A. Sheldon, R. S. Downing, Downing, Heterogeneous catalytic transformations for environmentally friendly production[J]. *Appl. Catal. A*, 1999, 189: 163.

[17] Notari B. Titanium silicalites[J]. *Catal. Today*, 1993, 18: 163.

[18] Lin M, Shu X,Wang X, et al. Titanium-silicalite molecular sieve and the method for its preparation [P], US 6475465[P], 2002.

[19] Wang Y, Tuel A. Nanoporous zeolite single crystals: ZSM-5 nanoboxes with uniform intracrystalline hollow structures[J]. *MicroporousMesoporous Mater.*, 2008, 113(1): 286–295.

[20] Zuo Y, Wang X, Guo X. Synthesis of titanium silicalite-1 with small crystal size by using mother liquid of titanium silicalite-1 as seed[J]. *Ind. Eng. Chem. Res.*, 2011, 50(14): 8485–8491.

[21] Wang Y, Lin M, Tuel A. Hollow TS-1 crystals formed via a dissolution–recrystallization process[J]. *MicroporousMesoporous Mater.*, 2007, 102: 80.

[22] Zheng A, Xia C, Xiang Y, et al. Titanium species in deactivated HTS-1 zeolite from industrial cyclohexanone ammoxidation process[J]. *Catal. Commun.*, 2014, 45: 34.

[23] Mu X, Wang D, Wang Y, et al. Nanosized molecular sieves as petroleum refining and petrochemical catalysts[J]. *Chin. J. Catal.*, 2013, 34(1): 69–79.

[24] Sun B, Zhu L, Study on ammoximation of cyclohexanone to cyclohexanone oxime catalyzed by titanium silicalite-1 zeolite[J]. *China Pet. Process. Petrochem.*, 2001, 32: 22.

[25] Li Y, Wu W, Min E, et al. Intrinsic kinetics of ammoximation of cyclohexanone to cyclohexanone oxime over HTS zeolite[J]. *China Pet. Process. Petrochem.*, 2003, 34: 39.

[26] Bellussi G, Millini R, Pollesel P, et al. Zeolite science and technology at Eni[J]. *New J. Chem.*, 2016, 40, 4061–4077.

[27] Thangaraj A, Kumar R, Ratnasamy P. Catalytic properties of crystalline titanium silicalites I. Synthesis and characterization of titanium-rich zeolites with MFI structure[J]. *J. Catal.*, 1991, 131(1): 294–297.

[28] Thangaraj A, Eapen M J, Sivasanker S, et al. Studies on the synthesis of titanium silicalite, TS-1[J]. *Zeolites*, 1992, 12(8): 943–950.

[29] Peregot G, Bellussi G, Corno C, et al. Titanium-silicalite: a novel derivative in the pentasil family[J]. *Stud. Surf. Sci. Catal.*, 1986, 28: 129–136.

[30] Zhou J, Hua Z, Cui X, et al. Hierarchical mesoporous TS-1 zeolite: a highly active and extraordinarily stable catalyst for the selective oxidation of 2, 3, 6-trimethylphenol[J]. *Chem. Commun.*, 2010, 46(27): 4994–4996.

[31] Xin H, Zhao J, Xu S, et al. Enhanced catalytic oxidation by hierarchically structured TS-1 zeolite [J]. *J. Phys. Chem. C*, 2010, 114(14): 6553–6559.

[32] Corma A, Navarro M T, Pariente J P. Synthesis of an ultralarge pore titanium silicate isomorphous to MCM-41 and its application as a catalyst for selective oxidation of hydrocarbons [J]. *J. Chem. Soc. Chem. Commun.*, 1994, 2, 147–148.

[33] Wang W, Li G, Li W, Liu L. Synthesis of hierarchical TS-1 by caramel templating [J]. *Chem. Commun.*, 2011, 47: 3529–3531.

[34] Groen J C, Peffer L A A, Pérez-Ramírez J. Pore size determination in modified micro- and mesoporous materials. *Microporous Mesoporous Mater.*, 2003, 60(1): 1–17.

[35] Groen J C, Jansen J C, Moulijn J A, et al. Optimal aluminum-assisted mesoporosity development in MFI zeolites by desilication [J]. *J. Phys. Chem. B*, 2004, 108(35): 13062–13065.

[36] Friedrich H, de Jongh P E, Verkleij A J, et al. Electron tomography for heterogeneous catalysts and related nanostructured materials [J]. *Chem. Rev.*, 2009, 109(5): 1613–1629.

[37] Janssen A H, Koster A J, de Jong K P. Three - dimensional transmission electron microscopic observations of mesopores in dealuminated zeolite Y [J]. *Angew. Chem. Int. Ed.*, 2001, 113(6): 1136–1138.

[38] Weyland M, Midgley P A, Thomas J M. Electron tomography of nanoparticle catalysts on porous supports: A new technique based on Rutherford scattering [J]. *J. Phys. Chem. B*, 2001, 105(33): 7882–7886.

[39] On D T, Bonneviot L, Bittar A, et al. Titanium sites in titanium silicalites: an XPS, XANES and EXAFS study [J]. XPS characterization of a synthetic Ti - containing MFI zeolite framework: the titanosilicalites, TS-1 [J]. *J. Mol. Catal.*, 1992, 74(1): 233–246.

[40] Langerame F, Salvi A M, Silletti M, et al. XPS characterization of a synthetic Ti-containing MFI zeolite framework: the titanosilicalites, TS-1 [J]. *Surf. Interf. Anal.*, 2008, 40(3–4): 695–699.

[41] Camblor M A, Corma A, Perez-Pariente J. Infrared spectroscopic investigation of titanium in zeolites. A new assignment of the 960 cm^{-1} band [J]. *J. Chem. Soc. Chem. Commun.*, 1993 6, 557–559.

[42] Huang D G, Zhang X, Liu T W, et al. Synthesis of high-performanced titanium silicalite-1 zeolite at very low usage of tetrapropyl ammonium hydroxide [J]. Synthesis of high-performanced titanium silicalite-1 zeolite at very low usage of tetrapropyl ammonium hydroxide [J]. *Ind. Eng. Chem. Res.*, 2013, 52(10): 3762–3772.

[43] Tuel A, Taarit Y B. Variable temperature ^{29}Si MAS NMR studies of titanosilicalite TS-1 [J]. *J. Chem. Soc. Chem. Commun.*, 1992 21, 1578–1580.

[44] Zhao Q, Bao X H, Han X W, et al.Studies on the crystallization process of titanium silicalite-1 (TS-1) synthesized using tetrapropylammonium bromide as a template [J]. *Mater. Chem. Phys.*, 2000, 66(1): 41–50.

[45] Fan W, Duan R G, Yokoi T, et al. Synthesis, crystallization mechanism, and catalytic properties of titanium-rich TS-1 free of extraframework titanium species [J]. *J. Am. Chem. Soc.*, 2008, 130(31): 10150–10164.

[46] Mielby J, Abildstrøm J O, Wang F, et al. Oxidation of Bioethanol using Zeolite - Encapsulated Gold Nanoparticles [J]. *Angew. Chem. Int. Ed.*, 2014, 126(46): 12721–12724.

[47] Li S, Burel L, Aquino C, et al. Transition-Metal Nanoparticles in Hollow Zeolite Single Crystals as Bifunctional and Size-Selective Hydrogenation Catalysts [J]. *Chem. Commun.*, 2013, 49(76): 8507–8509.

[48] Li S, Boucheron T, Tuel A, et al. Size-selective hydrogenation at the subnanometer scale over platinum nanoparticles encapsulated in silicalite-1 single crystal hollow shells [J]. *Chem. Commun.*, 2014, 50(15): 1824–1826.

[49] Ivanova I I, Knyazeva E E. Micro–mesoporous materials obtained by zeolite recrystallization: synthesis, characterization and catalytic applications [J]. *Chem. Soc. Rev.*, 2013, 42(9): 3671–3688.

[50] Ivanova I I, Kasyanov I A, Maerle A A, et al. Mechanistic study of zeolites recrystallization into micro-mesoporous materials [J]. *Microporous Mesoporous Mater.*, 2014, 189: 163–172.

[51] Kasyanov I A, Maerle A A, Ivanova I I, et al. Towards understanding of the mechanism of stepwise zeolite recrystallization into micro/mesoporous materials [J]. *J. Mater. Chem. A*, 2014, 2(40): 16978–16988.

[52] Dal Pozzo L, Fornasari G, Monti T. TS-1, catalytic mechanism in cyclohexanone oxime production [J]. *Catal. Commun.*, 2002, 3: 369.

[53] Pinelli D, Trifirò F, Vaccari A, Giamello E, Pedulli G. Nature of active sites in catalytic ammoximation of cyclohexanone to the corresponding oxime on amorphous silica: EPR investigations [J]. *Catal. Lett.*, 1992, 13: 21.

[54] Anilkumar M, Höelderich W F. New non-zeolitic Nb-based catalysts for the gas-phase Beckmann rearrangement of cyclohexanone oxime to caprolactam [J]. *J. Catal.*, 2012, 293: 76.

[55] Xu L, Ding J, Yang Y, et al. Distinctions of hydroxylamine formation and decomposition in cyclohexanone ammoximation over microporous titanosilicates [J]. *J. Catal.*, 2014, 309: 1–10.

[56] Lin J, Xin F, Yang L, et al. Synthesis, characterization of hierarchical TS-1 and its catalytic performance for cyclohexanone ammoximation [J]. *Catal. Commun.*, 2014, 45: 104–108.

[57] Zhuo Z, Wang L, Zhang X, et al. Insights into the key to highly selective synthesis of oxime via ammoximation over titanosilicates [J]. *J. Catal.*, 2015, 329: 107–118.

[58] Dong C, Wang K, Zhang J S, et al. Reaction kinetics of cyclohexanone ammoximation over TS-1 catalyst in a microreactor [J]. *Chem. Eng. Sci.*, 2015, 126: 633–640.

[59] Zhuo Z, Wu L, Wang L, et al. Lewis acidic strength controlled highly selective synthesis of oxime via liquid-phase ammoximation over titanosilicates [J]. *RSC Adv.*, 2014, 4(99): 55685–55688.

[60] Ding J, Wu P. Selective synthesis of dimethyl ketone oxime through ammoximation over Ti-MOR catalyst [J]. *Appl. Catal. A Gen.*, 2014, 488: 86–95.

[61] Clerici M G. The activity of titanium silicalite-1 (TS-1): Some considerations on its origin [J]. *Kinet. Catal.*, 2015, 56(4): 450–455.

[62] PetriniG, Cesana A, De Alberti G, et al. Deactivation phenomena on Ti-silicalite [J]. *Stud. Surf. Sci. Catal.*, 1991, 69: 761.

[63] Grosch G H, Muller U, Walch A, et al., Method for regenerating a zeolitic catalyst[P] US 6710002[P], 2004.

[64] Clerici M G. Zeolites for fine chemicals production [J]. *Top. Catal.*, 2000, 13: 373.

[65] Sinclair P E, Catlow C R A. Catlow, Quantum chemical study of the mechanism of partial oxidation reactivity in titanosilicate catalysts: active site formation, oxygen transfer, and catalyst deactivation[J]. *J. Phys. Chem. B*, 1999, 103: 1084.

[66] Varagnat J. Hydroquinone and pyrocatechol production by direct oxidation of phenol [J]. *Ind. Eng. Chem. Prod. Res. Dev.*, 1976, 15(3): 212–215.

[67] Neidig M L, Hirsekorn K F. Hirsekorn. Insight into contributions to phenol selectivity in the solution oxidation of benzene to phenol with H_2O_2[J]. *Catal. Commun.*, 2011, 12(6): 480–484.

[68] Liang X, Yang R, Li G, et al. Phenol hydroxylation over Fe-incorporated mesoporous materials prepared by coprecipitation [J]. *Microporous Mesoporous Mater.*, 2013, 182: 62–72.

[69] Bassler P, Weidenbach M, Goebbel H. The new HPPO Process for Propylene Oxide: From Joint Development to World scale Production [J]. *Chem. Eng. Trans.*, 2010, 21: 571–576.

[70] Hayashi T, Tanaka K, Haruta M. Selective vapor-phase epoxidation of propylene over Au/TiO2 catalysts in the presence of oxygen and hydrogen [J]. *J. Catal.*, 1998, 178(2): 566–575.

[71] Wu G, Wang Y, Wang L, et al. Epoxidation of propylene with H_2O_2 catalyzed by supported TS-1 catalyst in a fixed-bed reactor: experiments and kinetics [J]. *Chem. Eng. J.*, 2013, 215: 306–314.

[72] Russo V, Tesser R, Santacesaria E, et al. Chemical and technical aspects of propene oxide production via hydrogen peroxide (HPPO process) [J]. *Ind. Eng. Chem. Res.*, 2013, 52(3): 1168–1178.

[73] Climent M J, Corma A, Iborra S. Heterogeneous catalysts for the one-pot synthesis of chemicals and fine chemicals [J]. *Chem. Rev.*, 2011, 111(2): 1072–1133.

[74] http://corporate.evonik.com/en/company/locations/location-asia/pages/skc-evonik-peroxide-korea-co-ltd.aspx

[75] http://www.Chem.s-technology.com/projects/basf-hppo/

[76] http://corporate.evonik.com/en/media/press_releases/Pages/news-details.aspx?newsid=45021

[77] http://newsroom.dow.com/press-release/company-news/scg-dow-group-achieves-successful-thailand-hppo-plant-start

[78] Lin M, Xia C, Zhu B, et al. Green and efficient epoxidation of propylene with hydrogen peroxide (HPPO process) catalyzed by hollow TS-1 zeolite: A 1.0 kt/a pilot-scale study[J]. *Chem. Eng. J.*, 2016, 295: 370–375.

[79] Jiménez S C, Ruiz J R. The Baeyer–Villiger reaction on heterogeneous catalysts[J]. *Tetrahedron*, 2008, 64: 2011.

[80] Krow G R. *Org. React.*, 2004, 43: 251.

[81] Strukul G. Transition metal catalysis in the Baeyer–Villiger oxidation of ketones[J]. *Angew. Chem. Int. Ed.*, 1998, 37: 1198.

[82] Kotlewska A J, Rantwijk F, Sheldon R A, Arends I W C E. Epoxidation and Baeyer–Villiger oxidation using hydrogen peroxide and a lipase dissolved in ionic liquids [J]. *Green Chem.*, 2011, 13: 2154.

[83] Watanabe A, Uchida T, Ito K, Katsuki T, Highly enantioselective Baeyer–Villiger oxidation using Zr (salen) complex as catalyst [J]. *Tetrahedron Lett.*, 2002, 43: 4481.

[84] Xia C, Lin M, Bin Z, et al. Study on the Mechanisms for Baeyer-Villiger Oxidation of Cyclohexanone with Hydrogen Peroxide in Different Systems [J]. *China Pet. Process. Petrochem.*, 2012, 2: 004.

[85] Xia C, Bin Z, Min L, et al. "Green" Cyclohexanone Oxidation Route Catalyzed by Hollow Titanium Silicate Zeolite for Preparing ε-Caprolactone, 6-Hydroxyhexanoic Acid and Adipic Acid [J]. *China Pet. Process. Petrochem.* Technology, 2012, 4: 012.

[86] Xia C, Ju L, Zhao Y, et al. Heterogeneous oxidation of cyclohexanone catalyzed by TS-1: Combined experimental and DFT studies [J]. *Chin. J. Catal.*, 2015, 36(6): 845–854.

[87] Renz M, Blasco T, Corma A, Fornés V, Jensen R, Nemeth L. Selective and Shape-Selective Baeyer-Villiger Oxidations of Aromatic Aldehydes and Cyclic Ketones with Sn - Beta Zeolites and H_2O_2 [J]. *Chem. Eur. J.*, 2002, 8: 4708.

[88] Corma A, Nemeth L T, Renz M, Valencia S. Sn-zeolite beta as a heterogeneous chemoselective catalyst for Baeyer–Villiger oxidations [J]. *Nature*, 2001, 412: 423.

[89] Corma A, Navarro M T, Renz M. Lewis acidic Sn (IV) centers—grafted onto MCM-41—as catalytic sites for the Baeyer–Villiger oxidation with hydrogen peroxide[J]. *J. Catal.*, 2003, 219: 242.

[90] Boronat M, Concepción P, Corma A, Renz M. Peculiarities of Sn-Beta and potential industrial applications [J]. *Catal. Today*, 2007, 121: 39.

[91] Schuchardt U, Cardoso D, Sercheli R, et al. Cyclohexane oxidation continues to be a challenge [J]. *Appl. Catal. A Gen.*, 2001, 211(1): 1–17.

[92] Raja R, Sankar G, Thomas J M. Powerful redox molecular sieve catalysts for the selective oxidation of cyclohexane in air[J]. *J. Am. Chem. Soc.*, 1999, 121(50): 11926–11927.

[93] Hereijgers B P C, Weckhuysen B M. Aerobic oxidation of cyclohexane by gold-based catalysts: New mechanistic insight by thorough product analysis [J]. *J. Catal.*, 2010, 270(1): 16–25.

[94] Anand R, Hamdy M S, Gkourgkoulas P, et al. Liquid phase oxidation of cyclohexane over transition metal incorporated amorphous 3D-mesoporous silicates M-TUD-1 (M= Ti, Fe, Co and Cr) [J]. *Catal. Today*, 2006, 117(1): 279–283.

[95] Shi C, Zhu B, Lin M, et al. Cyclohexane mild oxidation catalyzed by new titanosilicate with hollow structure[J]. *Catal. Today*, 2011, 175(1): 398–403.

[96] Shi C, Zhu B, Lin M, et al. Synthesis and characterization of NTS material and its oxidation desulfurization properties[J]. *Catal. Today*, 2010, 149(1): 132–137.

[97] Koekkoek A J J, Kim W, Degirmenci V, et al. Catalytic performance of sheet-like Fe/ZSM-5 zeolites for the selective oxidation of benzene with nitrous oxide [J]. *J. Catal.*, 2013, 299: 81–89.

[98] Karthikeyan S, Gupta V K, Boopathy R, et al. A new approach for the degradation of high concentration of aromatic amine by heterocatalytic Fenton oxidation: kinetic and spectroscopic studies [J]. *J. Mol. Liq.*, 2012, 173: 153–163.

[99] Yuan C, Liang Y, Hernandez T, et al. Metal-free oxidation of aromatic carbon-hydrogen bonds through a reverse-rebound mechanism [J]. *Nature*, 2013, 499(7457): 192–196.

[100] Yuan C, Gao X, Pan Z, et al. Molybdovanadophosphoric anion ionic liquid as a reusable catalyst for solvent-free benzene oxidation to phenol by H_2O_2[J]. *Catal. Comm.*, 2015, 58: 215–218.

[101] Geng X, Lin M, Zhu B, et al. Method for preparing hydroquinone [P], CN101759530 B [P], 2008.

[102] Peng X, Lin M, Zhu B, et al. Method for preparing methylphenol by hydroxylating methylbenzene [P], CN201210129698.8[P], 2012.

Chapter 7

Luminescent Zeolites

Hui Lin and Minoru Fujii

Additional information is available at the end of the chapter

http://dx.doi.org/10.5772/63805

Abstract

Compared with the intensive fundamental research and practical application development of zeolites in the fields of photocatalysis, separation, ion exchange, and antimicrobial, research on the luminescence from optically functionalized zeolites is still at a juvenile stage. So far, plenty of important potential applications based on luminescent zeolites, such as light harvesting, phosphors, optical encoding/recording, micro-lasers, chemical sensors, and bio-imaging, have been demonstrated. What's more, it has been shown that the porous crystalline nature of zeolites makes them an excellent matrix for investigating the fundamental physics of the sub-nm-scale luminescent metal clusters, which is still not sufficiently clear till now. Due to its significance, luminescent zeolites functionalized by luminescence-active guest species in various types including organic dye molecules, rare earth (RE) ions, bismuth ions, and d^{10} metal ions are reviewed in this chapter, dedicating to presenting some representative work on luminescent zeolites during the past few decades, from materials design, preparation, to application demonstrations, in hope to boost more research interest and strength in this community.

To date, optically activated luminescent zeolites are mainly focused on several types of aluminosilicate zeolites with the dimension of pores or channels in the scale of about 0.3–1.5 nm, such as the FAU-type, the LTL-type, the LTA-type, the ZSM-5-type, and the SOD-type zeolites. Here, in this chapter, the zeolites mainly refer to these nanoporous zeolites functionalized by luminescent centers.

Keywords: luminescence, zeolites, rare earth ions, bismuth ions, noble and coinage metal clusters

1. Introduction

In Section 1, luminescence of zeolites functionalized by organic dye molecules and quantum dots incorporated within the cavities is reviewed. The strong capacity of incorporating and confining organic dye molecules and quantum dots into zeolites' cavities is enabled by the large inner surface area, the exchangeable extra-framework cations, the built-in static electrical field between the extra-framework cations and the framework oxygen, and the well-confined periodically distributed cavities. Previous work on organic dye molecules or quantum dots containing zeolites with designed alignment of the guest dye molecules and sometimes the host zeolite micro-/nanocrystals is briefly introduced. Since most of these important work has been well summarized in some previous book chapters, reviews, or feature articles (Gion Calzaferri, *Top Catal*, 2010, 53:130–140; Gion Calzaferri, *Chapter 7* in *Photofunctional zeolites: synthesis, characterization, photocatalysis reactions, light harvesting*; & Calzaferri, *Langmuir*, 2012, 28:6216–6231; Gion Calzaferri, *Advances in Photochemistry*, 2002, 27:1–50; Thomas Bein, *Chapter 18* in *Introduction to Zeolite Molecular Sieves*, 3rd Edition, 2007), here we generally review some typical methods for processing the dye molecules containing zeolites and introduce some new achievements in this area within the recent years.

The extra-framework cations of zeolites such as Na^+, K^+, and Ca^{2+} are not as stable as the framework Si^{4+} and Al^{3+} and are exchangeable to other cations, for instance, the usage of zeolite 4A as water softener via the $Na^+ \rightarrow Ca^{2+}$ or Mg^{2+} exchange. The extra-framework cations can also be replaced by optically active centers such as rare earth (RE) ions, the transition metal (TM) ions, and some luminescence-active main group elements ions, for instance, bismuth (Bi) ions, etc. The porous crystalline nature of zeolites benefits the homogeneous dispersion of luminescence-active centers in the zeolite matrices. To present efficient luminescence from the RE ion- and Bi ion-exchanged zeolites, relieving or eliminating the influence of the hydrated water molecules is a prerequisite. In Section 2, recent achievements on the visible and NIR emission from the rare earth (RE) ion- and bismuth (Bi) ion-exchanged zeolites and their potential optical and photonics applications are reviewed.

d^{10} ion-exchanged zeolites, especially Ag^+ ion-exchanged zeolites, show excellent photocatalysis capacity. Comprehensive work on summarizing the photocatalysis properties of transition metal ion-exchanged zeolites has been conducted in a previous book "Photofunctional Zeolites: Synthesis, Characterization, Photocatalytic Reactions, and Light Harvesting." In Section 3, we mainly focus on the luminescence properties of these materials, aiming for potential optical and photonics applications. Due to the closed-shell electronic structure, usually there exist weak metallophilic interactions between the d^{10} ions, which exhibit many intriguing luminescence phenomena. Amazingly, the capability of adsorption/desorption of small polar molecules of the aluminosilicate nanoporous zeolites can have strong influence on the coordination of the extra-framework d^{10} ions and the metallophilic interaction between the adjacent d^{10} ions; in result reversible emission variation from the d^{10} ion-exchanged zeolites has been observed.

1.1. Incorporating organic dye molecules into zeolites' cavities

Dye molecules incorporated within the cavities of zeolites are representative guest-host materials which have significance to both fundamental research and technical applications. The distinct feature of zeolite matrices is their nature-born, spatially perfectly confined cages and channels enabled by its porous crystalline structure, which makes them superior to porous glasses or dense crystalline matrices. Dye molecules are often encapsulated within zeolites' cavities via physisorption, self-assembly, ion exchange, crystallization inclusion, or direct synthesis in the cages or channels of zeolites (also called the ship-in-a-bottle method). The most frequently adopted zeolite matrices for organic dye molecules are those with strictly parallel one-dimensional (1D) channels, such as the LTL-type (short for L-type) and AFI-type zeolites, which can help simplify the investigation on the energy transfer process due to the strong 1D confinement of the dye molecules. By choosing proper organic dye molecules and zeolite matrix guest-host pairs, emission quenching due to the aggregation of dye molecules can be prevented by spatially confining the dye molecules at the monomer scale. Meantime, this features the high quantum yield emission by high concentration loading of organic dye molecules without aggregation-induced quenching. Anisotropic energy transfer among organic dye molecule guests and polarized emission parallel/perpendicular to the *c*-axis is possible to achieve within the dye molecules/1D-channel zeolites guest-host systems by proper spatial confinement of the dye molecules [1]. So far, various kinds of neutral, cationic organic dye molecules and their combination have been incorporated into the cavities of zeolites, including some dyes for laser generation [2]. To achieve patterned incorporation of the dye molecules into the orientation of well-aligned 1D-channel zeolite monolayers or arrays, sometimes micro-contact printing (mCP), which is usually used for fabricating or transferring the *c*-axis-oriented L-type or AFI-type zeolite monolayers or arrays, has been conducted [3].

Synthesis of *c*-axis-oriented 1D-channel zeolite monolayers or arrays with the channels open at the off-substrate ends and ready to load dye molecules has been of particular interest and intensively investigated [4]. To seal the entrance(s) to the 1D channels in case of dye molecule leakage after loading, stuffing at the end(s) of the 1D channels with molecules (phthalocyanine, for instance) serving as stopcock or surface coating is often performed [5]. In some cases, the stopcock molecules can also act as linkages so that they assemble the dye molecules containing 1D-channel zeolite micro-/nano-crystals end to end into chains.

1.2. Aligning the orientation of the 1D-channel zeolite micro-/nanocrystals in favor of photonics applications

Besides the controllable assembly of the organic dye molecules in the parallel 1D channels, orienting the micro-/nano-zeolite matrix grains to obtain well-aligned arrays or thin films is also highly expected to facilitate certain optical or photonic applications. Chemical modification [6], seeding-epitaxial growth [4], external magnetic field application [7], and micro-contact imprinting [8] have been utilized to align the zeolite micro-/nanocrystals along the 1D channel direction.

Obtaining zeolite micro-/nano-crystals monolayer via chemical interactions such as covalent linkages, ionic bonding, and hydrogen bonding has been summarized by Yoon [6]. Here we mainly introduce the orientation alignment via seeding-epitaxial growth and via external magnetic field assistance.

A typical process to obtain 1D-channel zeolite thin films with well-aligned grains by first loading a seeding layer on the substrate followed by the further epitaxial growth has been described as follows by Xue et al. [4]: first by surface modification of the quartz slice substrate using cation polymer, for example, poly(diallyldimethylammonium chloride) (PDDA), to get a positively charged substrate surface as template for the zeolite seed crystals; then seeds loading via immersing the substrate horizontally in the seed crystals suspension, followed by rinsing and drying to get the seeds slice; growth of the *c*-axis-oriented AFI film by a further hydrothermal crystallization in the autoclave; organic template (PDDA) removal by calcinations; and finally dye molecules loading by adding the calcined *c*-axis-oriented AFI thin film in an azo dye molecule (DPPA) solution. This azo dye molecule containing *c*-axis-oriented AFI zeolite film shows ability for detecting heavy metal ions by monitoring the optical absorption variation, particularly efficient for the detection of Zn^{2+} ions.

Cucinotta et al. depicted a simple experimental layout for aligning the orientation of the zeolite L micro-cylinders via applying external magnetic field [9]. After dye molecule loading, monodisperse Fe_3O_4 nanoparticles prepared by chemical co-precipitation were coated on the surface of zeolite L microcrystals via natural assembly without any surface modification. The attached Fe_3O_4 nanoparticles can assist the alignment and the orientation of the zeolite L microcrystals.

Amazingly, alignment and the orientation of organic dye molecules incorporated into zeolite L microcrystals decorated by the Fe_3O_4 nanoparticles via external magnetic field can be achieved in aqueous solution, on solid substrates, and even in polymer matrices. The most interesting point may be that anisotropy in the optical absorption and emission spectra was observed and assigned to the fine spatial alignment of both the hexagonal zeolite L microcrystal matrices and the organic dye molecules within the matrices' 1D channels.

1.3. Featured luminescence from the dye molecules incorporated into zeolites

The most intensively conducted work on the luminescence properties of organic dye molecules incorporated into 1D channel zeolites is regarding the design and fulfillment of the efficient energy transfer from the donors to acceptor dyes via the Förster resonant energy transfer (FRET), typically aiming for artificial photosynthesis. An extreme fast non-radiative energy transfer from the donor to the acceptor (emitter) is required in case of energy loss via other pathways [10].

It is interesting that some organic dyes, hostasol red (HR), for instance, show efficient luminescence in aprotic solvents but becomes inefficient when it's incorporated into zeolite L's channels. By partial or complete exchanging the extra-framework K^+ ions with imidazolium cation, the emission quantum yield of the hostasol red (HR) dyes containing zeolite L can be enhanced to a large extent by reducing the proton activity of the extra-framework cations in water molecules environment [11]. Meantime, the high-emission quantum yield can be together ensured by locating the HR dyes in the middle of the 1D channels and in the environment of perylene dye serving as the antenna donor, in which condition a highly efficient donor-acceptor system can be formed.

1.4. Quantum dots (clusters) embedded in zeolite matrices

Due to the tiny sizes of the nanoporous zeolites' cavities (0.3–1.5 nm), the sizes of the semiconductor nanospecies confined within the zeolites' cavities are smaller than typical semiconductor quantum dots and they can be at the scale which consists of several "molecules" [12]. At this scale, the tiny semiconductor crystals are sometimes called quantum clusters. Zeolite matrices have the advantages of excellent size confinement and periodical spacing between the quantum dots. Owing to the size well-defined cages and channels, zeolites have also been demonstrated for screening out the semiconductor dots or clusters with narrow size distribution.

1.5. Possible future work

At present, the applied magnetic field for aligning the orientation of zeolite microcrystals in the recent reports is very high (1 T). Actually, even for the paramagnetic materials, Al_2O_3, or $Ca_5(PO_4)_3F$, for instance, under such strong magnetic field, aligning grains along the *c*-axis can be also be achieved [13, 14]. Efforts may be deserved on providing other novel structures or architectures or decorating methods which can reduce the applied magnetic field to facilitate the orientation alignment with moderate magnetic field.

2. Rare earth (RE) ion- and bismuth (Bi) ion-exchanged nanoporous zeolites and their derivatives

2.1. Factors influencing the luminescence quantum efficiency and possible solutions

2.1.1. *Phonon energy of zeolites' frameworks*

The framework of aluminosilicate zeolites consists of $[SiO_4]^{4-}$ and $[AlO_4]^{5-}$ tetrahedrons sharing the O^{2-} corners; the lattice vibration energy of the (dehydrated) zeolite framework itself is suitable to serve as a host material for luminescence.

2.1.2. *Hydrate water molecules*

As mentioned above, for these zeolites with the aluminosilicate composition, to keep charge neutral, cations such as Na^+, K^+, and Ca^{2+} are located in the channels and cages and are coordinated by the framework oxygen. The electric field existing between the extra-framework cations and the framework oxygen has strong attraction for small polar molecules. As a result, water molecules are easily adsorbed in this type of zeolites such as the LTA and FAU types, usually with the H atoms bonded to the framework oxygen or with the O atoms coordinated to the extra-framework cations.

Hydrated water molecules are the most influencing factor on the emission efficiency of the optically functionalized luminescent zeolites. On the one hand, due to the high vibrational energy (maximal ~3600–3700 cm^{-1}), hydrated water molecules are generally detrimental to the emission efficiency of the RE and Bi ion-exchanged zeolites. To achieve efficient luminescence from these types of optically activated zeolites and its derivatives, dehydration and then preventing rehydration via stuffing the entrance for water molecules [15], or filling the cavities with molecules or radicals of low vibrational energy [16], are two often adopted methods.

On the other hand, interestingly, it has been demonstrated in our recent work that the hydrated water molecules are essential to the strong yellow-green emission from the Ag^+ ion-exchanged LTA zeolites, which can be attributed to the coordination to the Ag^+ ions that ensures an efficient ligand to metal charge transfer (LMCT) process. This part will be discussed with more details in Section 3.

2.2. Down-conversion near-infrared (NIR) emission from Nd^{3+} ion-exchanged zeolite Y

Nd^{3+} is a representative RE ion as the NIR luminescence-active center in various matrices (one peak at ~1.06 μm and the other at ~1.35 μm), for instance, the Nd^{3+}-doped phosphate glass or the Nd^{3+}-doped yttrium aluminum garnet (usually denoted as Nd^{3+}:YAG) single crystal for solid-state laser generation.

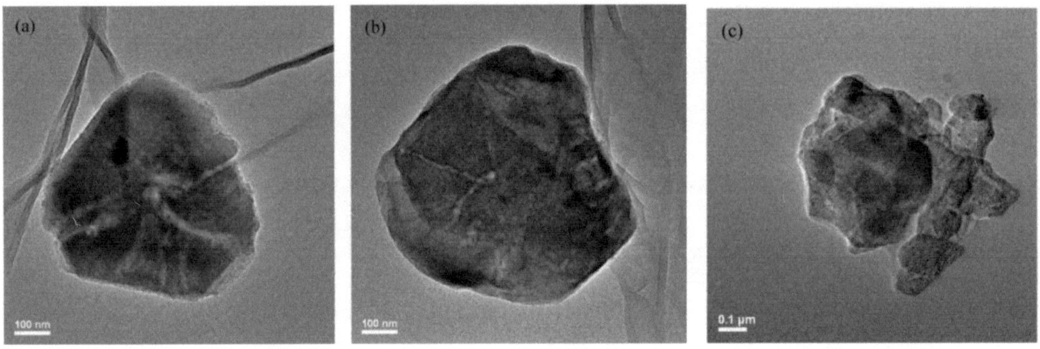

Figure 1. TEM morphology of pure zeolite Y (a), Nd^{3+} singly exchanged zeolite Y (b), and Nd^{3+}, Ag^+ ion-exchanged zeolite Y (c) individual particles treated at 900°C [17] (reproduced from American Institute of Physics).

As mentioned in Section 2.1.1, toward efficient emission from the RE or Bi ion-activated nanoporous zeolites, reducing or removing the hydrated water molecules is the key point. We found that co-loading Ag^+ ions into the cavities of zeolite Y can help collapse zeolite Y's structure; see **Figure 1**. Partial vitrification of the Nd^{3+} ion-exchanged zeolite Y was achieved at 900°C without extra crystalline phase introduction at the meantime. Partial elimination of water molecules can be identified from the FTIR spectra, where the absorption intensity due to the vibration of water molecules and hydroxyls both decreased for the partial vitrified sample, as shown in **Figure 2** (left). Due to the nonreversible dehydration in the vitrified parts of the partially collapsed Nd^{3+} ion-exchanged zeolite Y microcrystals, a six-fold enhancement of the 1053 nm NIR emission was achieved, as shown in **Figure 2** (right).

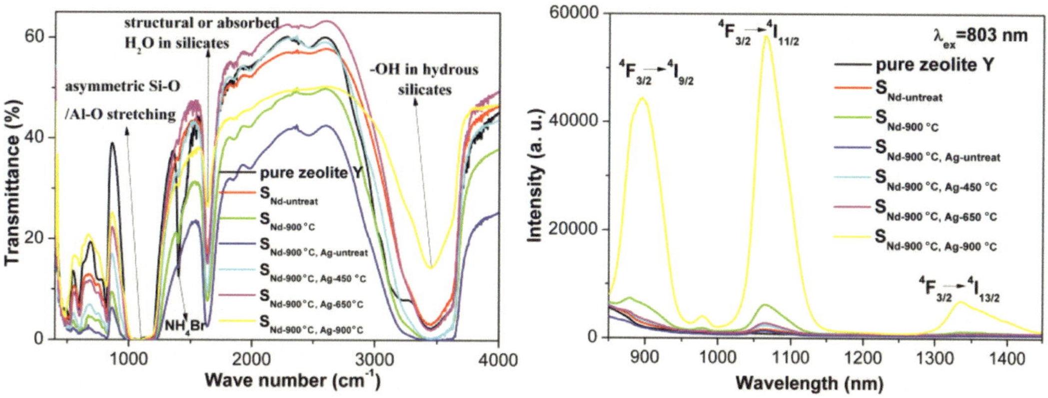

Figure 2. Left, FTIR spectra for the pure zeolite Y, Nd^{3+} ion-exchanged zeolite Y, and Nd^{3+}, Ag^{+} ion-exchanged zeolite Y under different thermal treatment conditions; right, NIR emission spectra of the Nd^{3+} (and Ag^{+}) ion-exchanged zeolite Y samples under different thermal treatment conditions [17] (reproduced from American Institute of Physics).

2.3. Up-conversion emission

Generally, there are few reports on the up-conversion (UC) emission from the RE ion-exchanged zeolites. This is probably due to the detrimental influence from hydrated water in the cavities, which can further decrease the low quantum efficiency of the up-conversion emission (usually less than 2%). Still, efforts have been devoted to this field, via extra coupling of molecules or radicals of low vibrational quanta, such as F^- and WO_4^{2-} [18], or partial dehydration by thermal treatment, or complete dehydration via collapsing the crystalline zeolite structure into glass-phase derivatives [19].

By employing the same method and taking advantage of blockade against water molecules by bismuth aggregates, that is, co-loading of bismuth ions into the zeolite Y's cavities combined with post-thermal treatment at 900°C, up-conversion green and red emissions have been achieved for the Er^{3+}, Yb^{3+}, and Bi^{3+} ions co-loaded zeolite Y powders [20].

2.4. Bismuth ion-exchanged zeolites for broad NIR emissions

2.4.1. Significance of the broad NIR emission from bismuth-activated materials

Since the first demonstration of broad NIR emission from bismuth-doped silica glass by Y. Fujimoto in 2001, research on the bismuth-activated broad NIR emitters has been conducted in various host lattices, involving nanocrystals, glasses, single crystals, and fibers (including both glass and crystalline fibers). The broad NIR (and mid-IR) emission from bismuth-activated luminescent materials falls well in the fiber-optic communication bands and hence is very promising as an alternative to the present commercial erbium (Er^{3+})- or thulium (Tm^{3+})-doped fibers and the erbium-doped fiber amplifiers (EDFA) for fiber-optic communication.

2.4.2. Broad NIR luminescence from bismuth ion-exchanged FAU zeolites and exploration on the NIR luminescent bismuth species

Exploration on the specific luminescent bismuth species is essential to promote the practical application of bismuth-doped materials, though it is complicated and challenging due to the rich oxidation states of bismuth ions. On the bismuth ion-exchanged zeolites, Sun and coworkers have contributed a series of work to the efficient broad NIR emission and the exploration on the bismuth species for the broad NIR emissions [15, 21, 22]. FAU zeolites, especially the Y-type possessing high-temperature structural duration due to the high Si/Al ratio, which is in favor of the dehydration process via relatively high-temperature thermal treatment, are chosen as the typical matrix for bismuth ions. Meantime, some by-product generated during the thermal treatment process may act as blockades for the Re-entry of water molecules. For instance, bismuth aggregates were proposed to be formed during the thermal annealing at 950°C and can prevent the water molecules from Re-entering the sodalite cages of zeolite Y; hence, strong NIR emission was achieved for the bismuth ion-exchanged zeolite Y powders, as depicted in **Figure 3**.

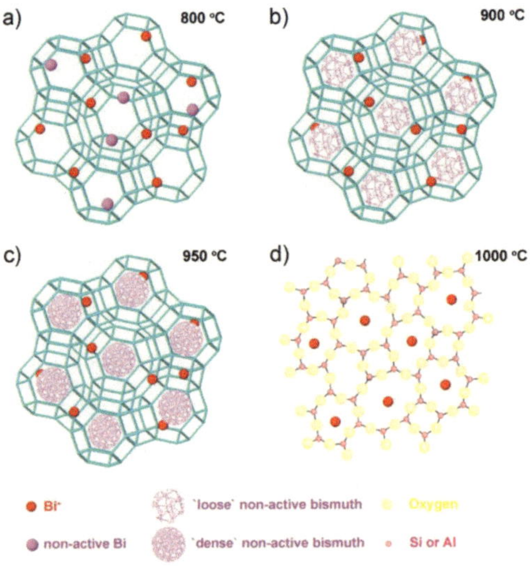

Figure 3. (a–c) Schematic illustration of the structure transformation of the zeolites annealed at different temperatures. (d) 2D representation of the structure of an aluminosilicate network. Note that "loose" and "dense" just represent small and large possibilities of sealing the pores of the zeolites with Bismuth agglomerates [15] (reproduced from John Wiley and Sons).

By high-resolution synchrotron X-ray diffraction (HR-XRD) (see **Figure 4**), photoluminescence (PL) spectra, electron spin resonance (ESR), nuclear magnetic resonance (NMR), and quantum chemistry calculation based on the spin-restricted relativistic time-dependent density functional theory (TDDFT), the broad emissions are assigned to several types of Bi^+ substructures in the sodalite cages of zeolite Y. Bi^+, Bi_2^{2+}, Bi_3^{3+}, and Bi_4^{4+} are proposed to be the active luminescence centers in the sodalite cages of zeolite Y. Specifically, NIR emissions at 1050 nm,

1135 nm, 1145 nm, and 1240/1285 nm are assigned to Bi^+, Bi_2^{2+}, Bi_3^{3+}, and Bi_4^{4+}, respectively (**Figure 5**).

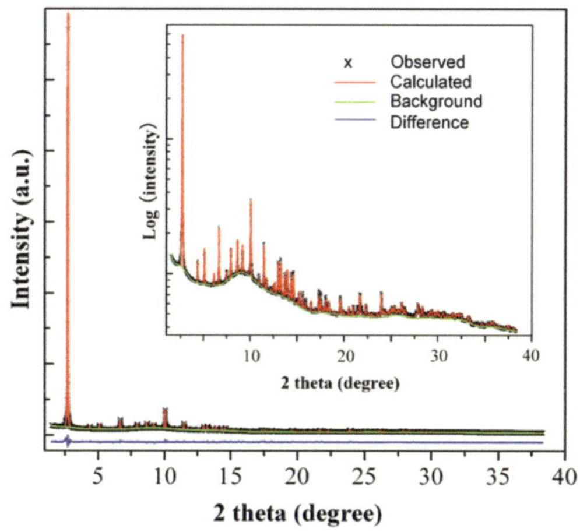

Figure 4. Rietveld fit to the high-resolution synchrotron XRD pattern of dehydrated Bi-embedded zeolite Y. Plots show the observed and calculated powder patterns. A difference curve is shown at the bottom of the diagram. Inset: a logarithmic-scale version of the graph [21] (reproduced from American Chemical Society).

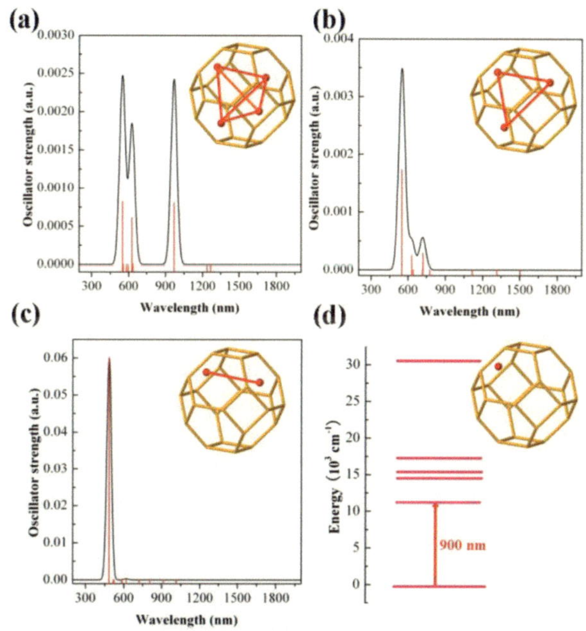

Figure 5. Theoretical absorption spectra for Bi_4^{4+} (a), Bi_3^{3+} (b), and Bi_2^{2+} (c). The energy-level diagram of Bi^+ (d) was drawn according to the experimental absorption spectrum of $AlCl_3$-NaCl molten salt containing Bi^+ [21] (reproduced from American Chemical Society).

2.5. Potential applications of the down- and up-conversion emission from RE and Bi ion-exchanged zeolites

Sun et al. demonstrated the bio-imaging in the second biological window with the amorphous aluminosilicate nanoparticles derived from FAU and LTA zeolites nanocrystals. The FAU and LTA zeolite nanocrystals are synthesized by the hydrothermal growth [23]. Bismuth ions were incorporated into the zeolite FAU and LTA nanocrystals' cavities via ion exchange and then subjected to thermal treatment at 870°C for 15 min, after which amorphous Bi ion-doped aluminosilicate nanoparticles were obtained. Finally, a 22 nm-thick silica layer was coated on the surface of Bi ion-doped aluminosilicate amorphous nanoparticles to improve the biocompatibility. In vivo bio-imaging in the second biological window (1–1.35 μm) for a mouse was successfully demonstrated by employing the silica-coated Bi ion-doped aluminosilicate nanoparticles derived from FAU and LTA zeolite nanocrystals.

Two main features of this material for in vivo bio-imaging applications are the relatively long emission decay time (hundreds of microseconds), which is in favor of the fluorescence lifetime imaging microscopy, and the emission wavelength in the second biological window, which provides higher spatial resolution than that of bio-imaging based on fluorescence in the first biological window (0.65–0.95 μm) [24].

Besides bio-imaging, down-conversion NIR emission from RE and Bi ion-exchanged zeolites which can split one UV-vis photon into two or more NIR photons may also have application as the converter layer to improve the conversion efficiency of crystalline silicon solar cells.

2.6. Possible future work

So far, average sizes of most commercial available zeolite powder matrices are in the range of micrometer or sub-micrometer scales; for instance, the typical edge length of the LTA zeolite cube (micro-single crystal) is about 1–2 μm. This size is too large for bio-imaging applications. Nanoporous zeolite nanocrystals with the average size from tens of, to several hundred nanometers, may be possible to get dispersed in water to facilitate the bio-applications.

Further detailed structure characterization of the RE and Bi ion-exchanged and thermal-treated zeolite micro-/nanocrystals is worth conducting, since the luminescent active centers incorporated in the glass-phase parts may also contribute to the emission.

It is worth trying whether it is possible to observe up-conversion emission color tuning via adsorption/de-adsorption of molecules, for example, small polar molecules such as H_2O, H_2S, and NH_3 or small organic molecules. Tuning of the UC emission color is expected to be enabled by the overall phonon energy change of the whole host-guest system upon adsorption/de-adsorption of different guest molecules of different phonon energies.

3. d^{10} ion-exchanged zeolites

3.1. One question about the reversible color change of Ag^+ ion-exchanged LTA zeolites

The reversible color change of Ag^+ ion-exchanged LTA was reported by Rálek et al. [25]. This color change was generally attributed to formation of Ag_m^{n+} clusters [26]. Till 1998, this phenomenon was explained in the way of charge transfer transition from the framework oxygen to the empty 5s orbit of the extra-framework Ag^+ ions [27]. This reversible color change can be achieved under very low Ag^+ ion loading concentration (Ag^+-Na^+ exchanging ratio equals 1/96) and dehydration at room temperature in vacuum without thermal treatment, in which condition, in principle Ag clusters were almost impossible to be formed due to the considerably light loading of Ag^+ ions. Interestingly, except for the assignment of the reversible color change, most of the later work on luminescence from the Ag^+ ion-exchanged zeolites still assigned the few-atom silver clusters [28–30], such as Ag_3^+ and Ag_6^+ to the luminescence origin other than the charge transfer transition [31, 32].

3.2. Methods to activate the Ag^+ ion-exchanged nanoporous luminescent zeolites

So far, various methods have been tried to activate the luminescence from the Ag^+ ion-exchanged nanoporous zeolites, such as UV laser illumination, X-ray irradiation, and the most often adopted thermal treatment. The role of UV laser in activating the visible emission can be observed by sharp contrast for one piece of Ag^+ ion-exchanged micro-crystal through comparing the regions with/without light irradiation. The mechanism of activation by UV illumination [33] and X-ray irradiation [34] was explained in the way of photo-reduction of Ag^+ into Ag^0 atoms and then the formation of Ag_n^{m+} clusters. It should be pointed out that besides the explanation from the view of formation of Ag_n^{m+} clusters, dehydration by the heat effect under the irradiation of high energy light beam may also contribute to the luminescence activation.

3.3. Reversible emission variation of Ag^+ ion-exchanged LTA and SOD zeolites and possible mechanisms for this process

It should be noted that the terminology "cluster" is widely adopted among different communities for different types of metal nanospecies. For instance, for metal-organic complexes, aggregates of metal ions in close contact via the weak metallophilic interaction are also called clusters (short for cluster A hereinafter). During the past two decades, few-atom, sub-nm scale (below 1 or 2 nm) noble and coinage (Au, Ag, Pt, Pd, and Cu) metal nanospecies bonded via the metal-metal bonds as the same as in bulk metals or molecular-like bonds are also called clusters (short for cluster B hereinafter).

So far, the origin of emission from cluster B is still not sufficiently clear. At present, explanations from the following two ways have been mainly proposed: one is based on the free electron theory combined with quantum confinement effect; the other is based on the super atom theory. To explore the emission mechanisms of the metal clusters, first the interaction type between the neighboring metal ions or atoms needs to be clarified. It is a pity that this issue did not seem to be clearly addressed in lots of the previous work.

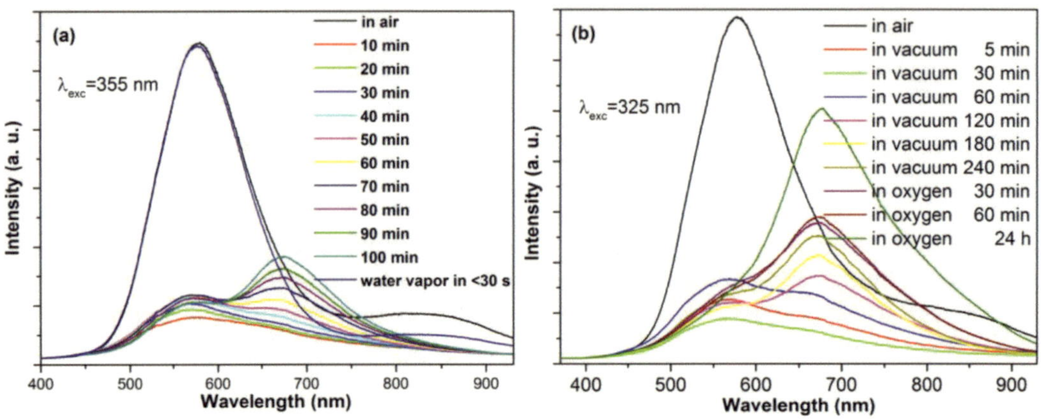

Figure 6. PL evolution of $Ag_{4.4}$, $Na_{7.6}$-A from air to vacuum then to water vapor (a) or oxygen (b) [35] (reproduced from American Institute of Physics).

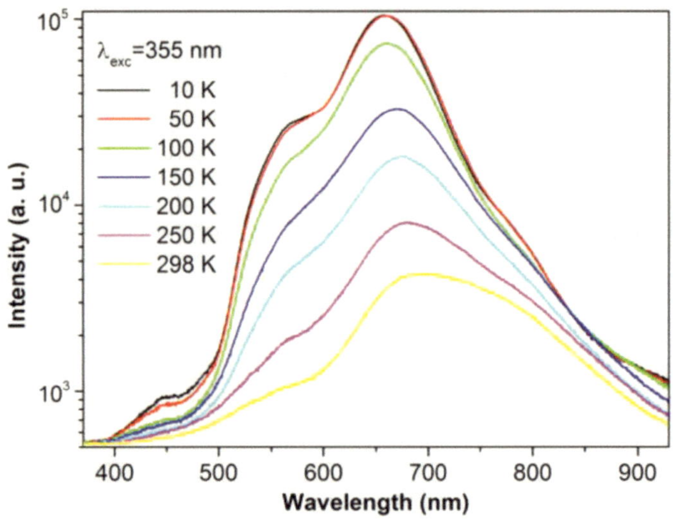

Figure 7. Temperature-dependent PL spectra for $Ag_{4.4}$, $Na_{7.6}$-A under the 355 nm excitation [35] (reproduced from American Institute of Physics).

Recently, H_2O molecule-sensitive, reversible green/red-dominant emission evolution was observed for the Ag^+ ion-exchanged LTA zeolite (see **Figure 6**), and different from most of the previous reports, the yellow-green emission is not assigned to Ag_n^{m+} cluster [35]. Confirmation of the role of H_2O molecules in the reversible emission evolution was achieved with the assistance of a designed vacuum chamber, into which different gas species such as N_2, O_2, and

H_2O vapor can be introduced. The reversible emission evolution process observed from the Ag^+ ion-exchanged LTA zeolites upon desorption/adsorption of H_2O molecules led us to reflect that the interaction between the neighboring Ag^+-Ag^+ ions does not seem to be the metal bonds while it may be the weak argentophilic (Ag^+---Ag^+) interaction.

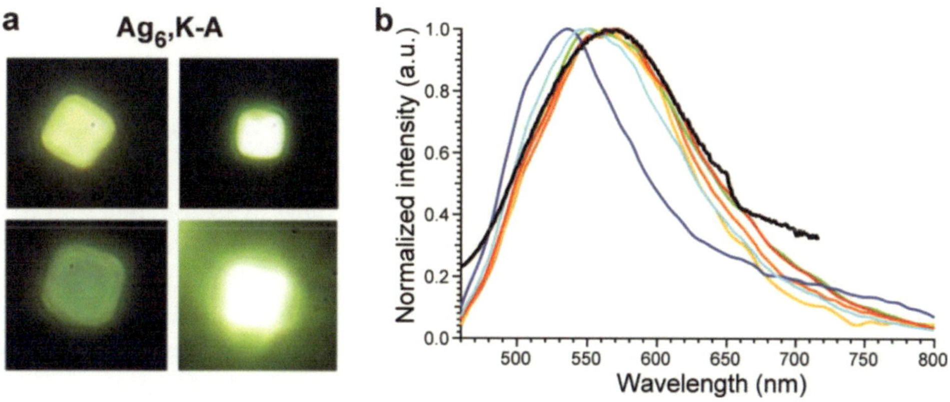

Figure 8. Single-crystal emission characterization for the Ag_3^+ clusters inside zeolite A upon UV excitation. (a) The individual crystals have sizes ranging from 2 to 5 μm. (b) Emission spectra of Ag_6,Na-A zeolites upon 400-nm illumination. The black curve corresponds to the bulk spectrum (measured in a fluorimeter), while the colored curves represent spectra from individual crystals measured on a confocal fluorescence microscope [28] (reproduced from the American Chemical Society).

There is no obvious narrowing of the emission band in the PL spectra of the Ag^+ ion-exchanged LTA zeolites taken at cryogenic temperature as low as 10 K (**Figure 7**). Pioneering work on comparing the PL spectra of the Ag^+ ion-exchanged LTA zeolite micro-crystals' ensemble and individual micro-crystals proves that there is no obvious inhomogeneous broadening contributing to the broad emission, as shown in **Figure 8**.

The H_2O molecule-sensitive, reversible emission evolution and the broad emission band at 10 K, together with the negligible contribution from inhomogeneous broadening, suggest that the emission from the Ag^+ ions containing LTA zeolites is not attributed to the quantum confinement effect. Corresponding to the origin of the reversible color change of Ag^+ ion-exchanged LTA zeolites upon dehydration/hydration, the green emission can be assigned to the framework $O^{2-} \rightarrow Ag^+$ (coordinated by water) ligand to metal charge transfer (LMCT) transition. Since there are two sets of cages (α-cages and β-cages) in LTA zeolites, it is difficult to propose further detailed physical scenario for the H_2O molecule-sensitive reversible green/red dominant emission evolution. So SOD zeolite, which consists of the sodalite cages and has very similar chemical composition to LTA zeolite, was employed as the matrix for Ag^+ ions, in hope to provide more specific information about the H_2O molecule-sensitive emission phenomenon.

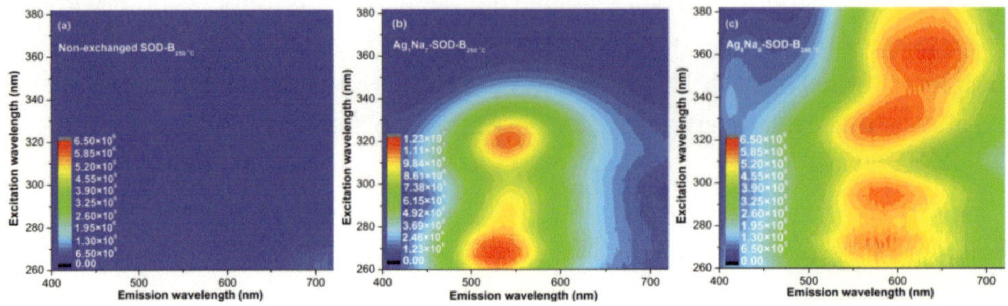

Figure 9. Excitation-emission 3D contours of non-exchanged SOD-$B_{250°C}$ (a), Ag_1Na_7-SOD-$B_{250°C}$ (b), and Ag_8Na_0-SOD-$B_{250°C}$ (c) [36] (reproduced from Royal Society of Chemistry).

The PL characteristics of the Ag^+ ion-exchanged SOD zeolites with different Ag^+ loading concentrations are similar to the Ag^+ ion-exchanged LTA counterparts. That is, under low Ag $^+$ loading concentration, only green emission can be observed, while under high Ag^+ loading concentration, red emission can be observed, as shown in **Figure 9**. Though not being as obvious as the situation for the Ag^+ ion-exchanged LTA zeolites, similar H_2O molecule-sensitive, reversible emission variation of the Ag^+ ion-exchanged SOD zeolites in vacuum/air or water vapor environment was observed [36], as shown in **Figure 10**.

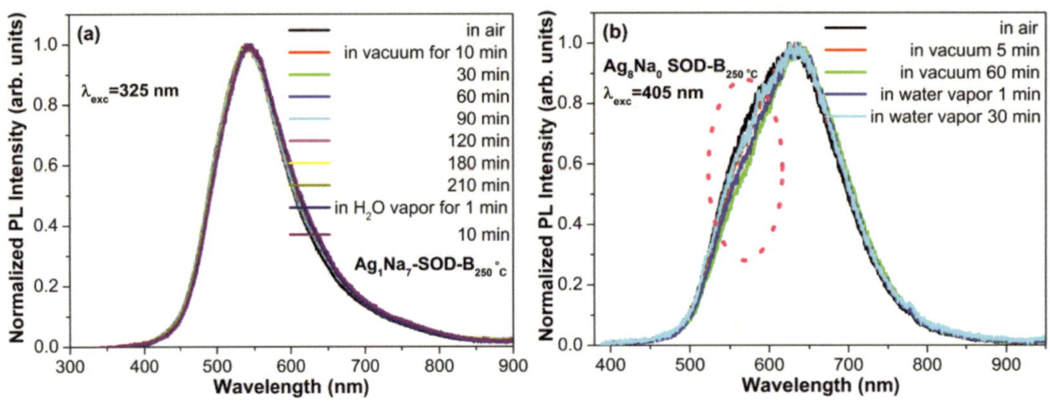

Figure 10. PL spectra recorded in different environmental conditions for Ag_1Na_7-SOD-$B_{250°C}$ (a) and Ag_8Na_0-SOD-$B_{250°C}$ (b) [38] (reproduced from Royal Society of Chemistry).

The low-lying excited states of Ag^+ ions (aggregates) enabling the visible or even NIR emission are commonly seen in the Ag(I)-organic complexes. In contrast, this phenomenon is rare for inorganic materials. So far, several types of inorganic materials have been reported with this kind of capability, such as the Ag^+-doped (or containing) oxyfluoride glasses, $K[Ag(I)CN_2]$ [37], or simply the Ag^+ ion-exchanged LTA and SOD zeolites. Note that for the Ag^+ ion-exchanged zeolites, it is shown that the amount of negative charges in zeolites' framework seems to have a key role in tuning down the emission energy, since sharp contrast has been observed between the emission energy of Ag^+ ion-exchanged LTA (or SOD) and Ag^+ ion-exchanged ZSM-5 zeolites. Specifically, green or red visible emissions can be observed from the Ag^+ ion-exchanged LTA (or SOD) zeolites (Al/Si ratio = 1), while only emissions in the UV region were observed from the Ag^+ ion-exchanged ZSM-5 zeolites (Al/Si ratio = 11.65). This is in good accordance with conclusions that the high-lying excited state of Ag^+ can be modified via ligand to metal or metal to metal charge transfer transition, especially when a reducing ligand is present [38].

With the complementary information obtained from the Ag^+ ion-exchanged SOD zeolites, a specific physical scenario for the H_2O molecule-sensitive reversible emission variation is proposed: the yellow-green emission is due to the ligand to metal (framework $O^{2-} \rightarrow Ag^+$) charge transfer transition within the partial hydration state, and the red emission is attributed to the ligand to metal-metal (framework $O^{2-} \rightarrow Ag^+$-Ag^+) or metal-metal (Ag^+-Ag^+) charge transfer transition upon dehydration when the water molecules which block the interaction between the neighboring Ag^+ ions are removed (see **Figure 11**).

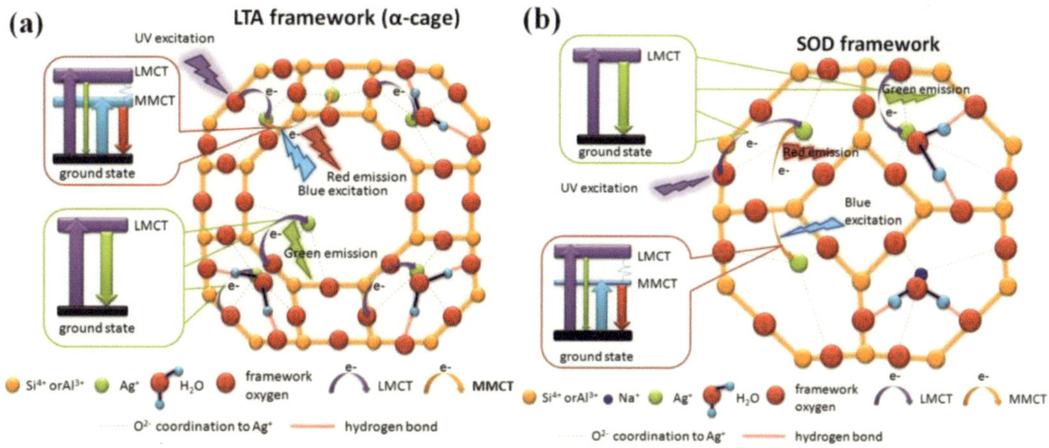

Figure 11. Proposed emission mechanisms in Ag^+ ion-exchanged LTA (a) and SOD (b) zeolites [36] (reproduced from Royal Society of Chemistry).

Though the emission variation tendency for the Ag^+ ion-exchanged SOD zeolites is similar to that of the Ag^+ ion-exchanged LTA zeolites, the recovery time for the emission variation from vacuum to H_2O vapor is much slower (~30 min) than the counterpart of Ag^+ ion-exchanged LTA zeolites (≤30 s). This can be attributed to the slow diffusion of H_2O molecules in SOD zeolites consisting of only the β-cages without straight, large channels (critical size 2.53 Å). In LTA, H_2O molecules can fast permeate along the 3D channels with the critical size of 4.21 Å confined by the 8-rings of the α-cage.

Location preference of the Ag^+ ions and Ag_n^{m+} clusters in LTA zeolite (either in the α-cage or the β-cage) is still controversial [39, 40]. From the contrasting response time of emission variation between the Ag^+ ion-exchanged LTA and SOD zeolites, it is suggested that for LTA zeolite Ag^+ ions are mainly located in the α-cage.

3.4. Potential optical and photonic applications of the Ag^+ ion-exchanged nanoporous zeolites

So far, by employing Ag^+ ion-exchanged nanoporous zeolites, promising optical and photonic applications have been demonstrated, such as optical recording and LED phosphors. Besides, the emission color change accompanied with reversible adsorption/de-adsorption of H_2O may have application as photoluminescence responsive sensors for small polar molecules or molecular tweezers.

Author details

Hui Lin and Minoru Fujii*

*Address all correspondence to: fujii@eedept.kobe-u.ac.jp

Department of Electrical and Electronic Engineering, Graduate School of Engineering, Kobe University, Kobe, Japan

References

[1] H. Lim, S.-E. Choi, H. Cheong, and J. S. Lee, *Micropor. Mesopor. Mater.*, 2014, 192, 89-94.

[2] L. Gartzia-Rivero, J. Banelos-Prieto, V. Martinez-Martinez, and I. Lopez Arbeloa, *ChemPlusChem*, 2012, 77, 61-70.

[3] M. Xue, L. Fan, Z. Kang, W. Zhang, H. Li, and S. Qiu, *J. Mater. Chem.*, 2012, 22, 17644-17648.

[4] N. S. Kehr, J. El-Gindi, H.-J. Gallab, and L. D. Cola, *Micropor. Mesopor. Mater.*, 2011, 144, 9-14.

[5] I. Lopez-Duarte, L.-Q. Dieu, I. Dolamic, M. V. Martinez-Diaz, T. Torres, G. Calzaferri, and D. Brühwiler, *Chem. Eur. J.*, 2011, 17, 1855-1862.

[6] K. Y. Yoon, *Acc. Chem. Res.*, 2007, 40, 29-40.

[7] S. Fibikar, G. Luppi, V. Martinez-Junza, M. Clemente-Leon, and L. D. Cola, *ChemPlusChem*, 2014, 79, 1-6.

[8] F. Cucinotta, Z. Popovic, E. A. Weiss, G. M. Whitesides, and L. D. Cola, *Adv. Mater.*, 2009, 21, 1142-1145.

[9] S. Fibikar, G. Luppi, V. Martínez-Junza, M. Clemente-León, and L. D. Cola, *ChemPlusChem*, 2015, 80(1), 62-67.

[10] G. Calzaferria, H. Maas, M. Pauchard, M. Pfenniger, S. Megelski, and A. Devaux, *Adv. Photochem.*, 2002, 27, 1-50.

[11] A. Devaux, G. Calzaferri, P. Belser, P. Cao, D. Brühwiler, and A. Kunzmann, *Chem. Mater.*, 2014, 26, 6878-6885.

[12] C. Leiggener and G. Calzaferri, *Chem. Eur. J.*, 2005, 11, 7191-7198.

[13] X. Mao, S. Wang, S. Shimai, and J. Guo, *J. Am. Ceram. Soc.*, 2008, 91(10), 3431-3433.

[14] J. Akiyama, Y. Sato, and T. Taira, *Opt. Lett.*, 2010, 35(21), 3598-3600.

[15] H.-T. Sun, A. Hosokawa, Y. Miwa, F. Shimaoka, M. Fujii, M. Mizuhata, S. Hayashi, and S. Deki, *Adv. Mater.*, 2009, 21, 3694-3698.

[16] Y. Wada, T. Okubo, M. Ryo, T. Nakazawa, Y. Hasegawa, and S. Yanagida, *J. Am. Chem. Soc.*, 2000, 122, 8583-8584.

[17] H. Lin, S. Rong Gui, K. Imakita, and M. Fujii, *J. Appl. Phys.*, 2014, 115, 033507.

[18] M. M. Lezhnina, and U. H. Kynast, *Phys. Sol. State*, 2005, 47, 1485-1488.

[19] Y. Gong, H. Chen, Q. He, J. Shi, L. Wang, W. Jiang, *Ceram. Int.*, 2013, 39, 8865-8868.

[20] Z. Bai, M. Fujii, K. Imakita, and S. Hayashi, *Micropor. Mesopor. Mater.*, 2013, 173, 43-46.

[21] H.-T. Sun, Y. Matsushita, Y. Sakka, N. Shirahata, M. Tanaka, Y. Katsuya, H. Gao, and K. Kobayashi, *J. Am. Chem. Soc.*, 2012, 134, 2918-2921.

[22] H.-T. Sun, Y. Sakka, N. Shirahata, Y. Matsushita, K. Deguchi, and T. Shimizu, *J. Phys. Chem. C*, 2013, 117, 6399-6408.

[23] H.-T. Sun, J. Yang, M. Fujii, Y. Sakka, Y. Zhu, T. Asahara, N. Shirahata, M. Ii, Z. Bai, J. Li, and H. Gao, *Small*, 2011, 7, 199-203.

[24] G. Hong, J. C. Lee, J. T. Robinson, U. Raaz, L. Xie, N. F. Huang, J. P. Cooke, and H. Dai, *Nat. Med.*, 18(12), 1841-1848 (2012).

[25] M. Rálek, P. Jírů, O. Grubner, and H. Beyer, *Collect. Czech. Chem. Commun.*, 1962, 27, 142.

[26] L. R. Gellens, W. J. Mortier, R. A. Schoonheydt, and J. B. Uytterhoeven, *J. Phys. Chem.*, 1981, 85, 2783.

[27] R. Seifert, A. Kunzmann, and G. Calzaferri, *Angew. Chem. Int. Ed.*, 1998, 37, 1521.

[28] G. D. Cremer, E. Coutiño-Gonzalez, M. B. J. Roeffaers, B. Moens, J. Ollevier, M. V. der Auweraer, R. Schoonheydt, P. A. Jacobs, F. C. De Schryver, J. Hofkens, D. E. De Vos, B. F. Sels, and T. Vosch, *J. Am. Chem. Soc.*, 2009, 131, 3049.

[29] H. Lin, K. Imakita, S.C.R. Gui, and M. Fujii, *J. Appl. Phys.*, 2014, 116, 013509.

[30] H. Hoshino, Y. Sannohe, Y. Suzuki, T. Azuhata, T. Miyanaga, K. Yaginuma, M. Itoh, T. Shigeno, Y. Osawa, and Y. Kimura, *J. Phys. Soc. Jpn.*, 2008, 77, 064712.

[31] G. Calzaferri, C. Leiggener, S. Glaus, D. Schürch, and K. Kuge, *Chem. Soc. Rev.*, 2003, 32, 29.

[32] R. Seifert, R. Rytz, and G. Calzaferri, *J. Phys. Chem. A*, 2000, 104, 7473.

[33] G. D. Cremer, Y. Antoku, M. B. J. Roeffaers, M. Sliwa, J. V. Noyen, S. Smout, J. Hofkens, D. E. De Vos, B. F. Sels, and T. Vosch, *Angew. Chem. Int. Ed.*, 2008, 47, 2813.

[34] E. Coutino-Gonzalez, D. Grandjean, M. Roeffaers, K. Kvashnina, E. Fron, B. Dieu, G. D. Cremer, P. Lievens, B. Sels, and J. Hofkens, *Chem. Commun.*, 2014, 50, 1350.

[35] H. Lin, K. Imakita, and M. Fujii, *Appl. Phys. Lett.*, 2014, 105, 211903.

[36] H. Lin, K. Imakita, M. Fujii, V. Y. Prokof'ev, N. E. Gordina, B. Saïdc, and A. Galarneau, *Nanoscale*, 2015, 7, 15665-15671.

[37] M. A. Rawashdeh-Omary, M. A. Omary, H. H. Patterson, and J. P. Fackler, Jr., *J. Am. Chem. Soc.*, 2001, 123(45), 11237-11247.

[38] F. Sabin, C. K. Ryu, P. C. Ford, and A. Vogler, *Inorg. Chem.*, 1992, 31(10), 1941-1945.

[39] R. Seifert, R. Rytz, and G. Calzaferri, *J. Phys. Chem. A*, 2000, 104, 7473.

[40] T. Sun and K. Seff, *Chem. Rev.*, 1994, 94, 857.

Chapter 8

XAFS and IR Studies on Luminescent Silver Zeolites

Takafumi Miyanaga, Yushi Suzuki and Sho Narita

Additional information is available at the end of the chapter

http://dx.doi.org/10.5772/63904

Abstract

In this chapter, studies of structures and optical properties of Ag-zeolite A by means of X-ray absorption fine structure (XAFS) and infrared (IR) spectra are presented. XAFS is a powerful tool to study the local structure of Ag nanoclusters in the zeolite cavity and IR spectra are quite sensitive to the change in zeolite lattice affected by the production of Ag clusters. First, we focus on the creation of Ag clusters in zeolite A by heat treatment under atmosphere and vacuum. Second, we discuss the mechanism of photoluminescence for Ag-zeolite composite. It is widely believed that the emitting point is Ag clusters in the Ag zeolite; on the other hand, our recent result is contradicted that Ag clusters are broken down in the strongly emitting species.

Keywords: silver zeolite A, Ag cluster, photoluminescence, XAFS, IR spectroscopy

1. Introduction

Ag zeolites are versatile family of functional materials: application to catalysis, anti-bacterial materials, information storage, pressure, and chemical sensors. In the field of nanoscaled science, the metal nanoparticles and nanoclusters are quite quite attractive. Using the zeolite cavity, nanoparticles and clusters are produced simply and the particle size is easily controlled [1–4]. However, the structural effect of zeolite lattice is not so well known that the structure and properties of nanoparticles and clusters are quite complicated.

Optical properties of small metal clusters are intriguing research topics today. Luminescence from the metal clusters is an attractive subject and especially silver and gold clusters show brilliant emission. This luminescent behavior can be expected by industry market without rare-earth metals. Recently, the characterization of photo luminescence from heat-treated Ag zeolites is extensively investigated [5–7]. So far, the luminescent species are widely believed to be Ag clusters or oligomers produced in the cavity of zeolite. However, there are still ambiguous points and hot discussions in the field of luminescent Ag zeolite.

X-ray absorption fine structure (XAFS) has characteristic features that elemental selectively analyzed with local structure for such complex systems. In the case of Ag zeolites, the local and electronic structure around Ag atoms has been extensively investigated [8–11]. On the other hand, infrared (IR) spectra are sensitive to molecular atomic vibration of zeolite lattice. In this chapter, we present the introduction of XAFS and IR spectroscopy first and then discuss (1) the creation of Ag clusters in the cavity of zeolite A and (2) the mechanism of photoluminescence (PL) of Ag-zeolite composites by the use of XAFS and IR spectra [12].

2. X-ray absorption fine structure and infrared spectra

In this section, we show the outline of the tools for the investigation of the structure and electronic states of Ag zeolites: XAFS and IR.

2.1. XAFS spectroscopy

We present the basic description of XAFS technique theoretically and experimentally. XAFS is the fine structure observed on X-ray absorption coefficient spectra near the absorption edge of a particular element. When monochromatic X-rays are irradiated on a sample, the intensity of the transmitted X-ray, I_t, depends on the photon energy E and sample thickness d (**Figure 1**). According to the Beer-Lambert law, I_t is expressed as

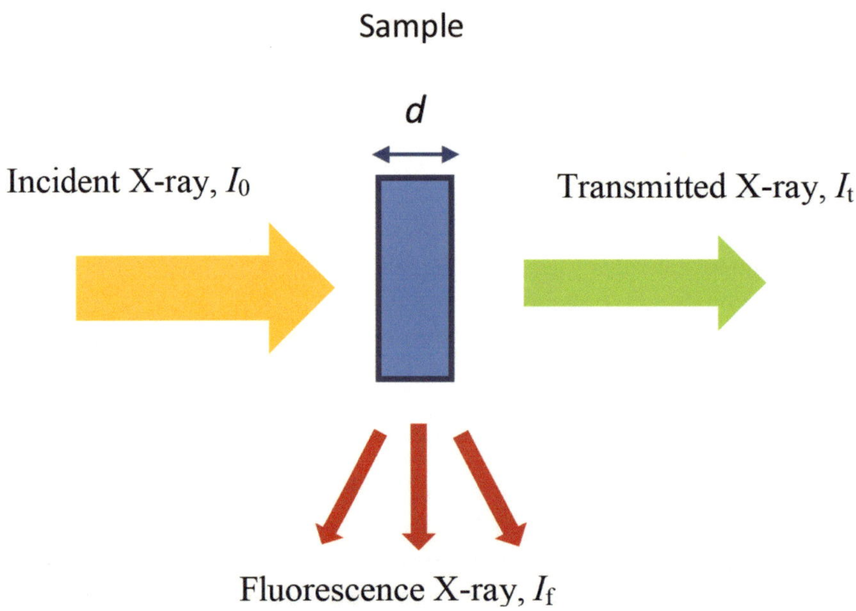

Figure 1. Schematic representation of transmission and fluorescence mode for X-ray absorption.

$$I_t = I_0 e^{-\mu(E)d} \quad (1)$$

Here, I_0 is the intensity of the incident X-ray and $\mu (E)$ is the energy-dependent X-ray absorption coefficient. Since $\mu(E)$ is approximately proportional to $a\ Z^4/m\ E^3$ where a, Z, and m are, respectively, the density of the sample, atomic number, and electron mass, the $\mu(E)$ decreases monotonically as a function of X-ray energy. When the X-ray energy has an exciting energy of a core electron, the core electron ejects to continuum state. In that time, the X-ray photon is annihilated in order to create a photoelectron and a core hole. As the result, $\mu(E)$ shows stepwise increase at the energy of core electron excitation (absorption edge). The core hole recombines with an electron from the higher energy level; X-ray fluorescence or the creation of an Auger electron is emitted.

In the X-ray energy region generally 60–1000 eV from the absorption edge, the equation of the absorption coefficient can be simplified under the following approximations as the Muffin-tin approximation for the atomic potential, the one-electron process in the absorption, the dipole approximation for the interaction between X-ray and matter, the plane-wave approximation for the propagating photoelectron, and the single scattering process considered in the scattering process. The X-ray absorption profile in this high-energy region is called Extended XAFS (EXAFS), where function $\chi(k)$ is expressed as

$$\chi(k) = \sum_j \frac{N_j S_0^2}{k r_j^2} \left| f_j(k,\pi) \right| e^{\left(-2\left(\sigma_j^2 k^2 + r_j/\lambda(k)\right)\right)} \sin\left(2kr_j + 2\delta_i(k) + \phi_j(k)\right), \quad (2)$$

where r_j, N_j, and σ_j are the interatomic distance between X-ray absorbing atom i and electron scattering atom j, the coordination number of scattering atom j, and the root-mean-square relative displacement (or Debye-Waller factor) of the atomic pair ij, respectively. These unknown structural parameters can be determined by the use of the nonlinear least-square-fitting (the so-called curve-fitting) method; for example, XANADU [13] code and Athena & Artemis code [14]. The phase shift of the photoelectron $\phi_j k)$, the electron backscattering amplitude $f_j(k)$, and the electron mean free path "lambda(symbol)" (k) in the medium are generally calculated from theoretically, for example, FEFF code [15].

The X-ray absorption coefficient can be determined from the ratio of intensities for the incident and transmitted X-ray as transmission mode, X-ray fluorescence as fluorescence mode, or Auger electrons as electron yield mode.

The intensity of incident and transmitted X-ray is measured using ionization chambers filled by inert gas. For transmission measurements, appropriate concentration is important for sample preparation. And then, the samples are required to be homogeneous and to have a constant thickness without pinholes. the samples are prepared by grinding and mixed with a suitable binder such as BN as transparent to X-rays in general.

We apply the transmission mode for the XAFS measurement of Ag zeolites. The Ag *K*-edge (25 keV) XAFS were measured at NW10A on PF-AR at KEK using Si(311) monochromator. We use in situ sample chamber shown in **Figure 2**, in which temperature can be controlled from LN_2 temperature to 773 K and the introduction of the various gasses is available.

Figure 2. In situ cell chamber for the measurement of Ag *K*-edge XAFS and IR spectra for Ag zeolites.

2.2. IR spectroscopy: basics and measurements

In this subsection, the method of infrared spectra is described. Vibrational spectroscopy has gained wide recognition for its ability to determine the chemical nature of materials. Three techniques have emerged to be most useful, namely Infrared Spectroscopy (IR), Electron Energy Loss Spectroscopy (EELS), and Raman Spectroscopy. Infrared spectroscopy has achieved high sensitivity at high resolution. Furthermore, this technique has proven to be most valuable due to its ability to work at ambient pressure from ultra-high vacuum (UHV) to atmosphere. EELS is restricted to UHV pressures and homogeneous (i.e., orders) surface. Since in the measurement of zeolite the coexistence gas is important, EELS is not suitable. Raman spectroscopy commonly used visible light as the excitation light, the energy is close to the PL excitation energy of Ag zeolite, and it would give a perturbation to the system. In addition, Raman spectroscopy has the disadvantage of sensitivity and is not suitable for the zeolite.

In XAFS, high-energy X-rays are used as probes, whereas low-energy infrared light is used in the IR as a probe. In the measurement of the Ag zeolite, XAFS detects the local structure around Ag atom but IR does that of the framework of zeolite. Thus, on measuring Ag zeolite, XAFS and IR can be said to be best partner for obtaining the complementary structural information.

The IR light absorption spectra were obtained at normal incidence of radiation. The infrared measurements were performed on an ABB BOMEM MB-100 spectrometer equipped with a DTGS detector at 4 cm^{-1} resolution with 128 interferometric scans. Zeolite powder was deposited on the Si wafer, and then an excessive specimen was blown off with a hand blower. A small quantity of zeolite was stuck electrostatically. The Si was installed with a sample into the in situ chamber cell as shown in **Figure 2**. The principle of signal collection is similar to the XAFS, as described in Eq. (1).

3. Sample preparation

The fully Ag^+-exchanged zeolite A was prepared by immersing the hydrated 12Na-A in 0.1-M $AgNO_3$ solution for 24 h at 25 °C. The solution was stirred every 1 h. After careful filtration, the Ag-zeolite-A powder was dried under air at room temperature (RT) in the dark space.

We apply two kinds of methods for producing Ag clusters in the cavity as follows:

(a) Ag-zeolite A was heated at 300–500 °C in vacuum to produce the Ag clusters [11]. After maintaining at high temperature for 24 h, it was cooled to room temperature, and then various gases, air, O_2, N_2, H_2O, and their mixtures, are introduced.

(b) Ag-zeolite A was heated at 300–500 °C under atmosphere and maintained at the same temperature for 24 h. After that, it is cooled to room temperature under atmosphere (with air).

For both processes, the samples were kept for long time (about 24 h) after cooling to room temperature.

4. Creation of Ag clusters in zeolite A cavity

In this section, we discuss the creation of Ag clusters in the zeolite cavity of zeolite A with heat treatment under atmosphere or vacuum studied by using XAFS.

Figure 3 shows the typical example of Fourier transform (FT) of the Ag *K*-edge EXAFS spectra for the hydrated Ag-zeolite A measured at RT atmosphere and after heating at 300 °C under vacuum. We can find two main peaks in FT: the first peak at around 2.2 Å and the second one at around 3.0 Å. The first peak is higher than the second one for the hydrated Ag-zeolite A, while the second peak is higher than the first one for the dehydrated Ag-zeolite A under vacuum. This phenomenon indicates that the Ag clusters are produced by heating at least 300°C under vacuum.

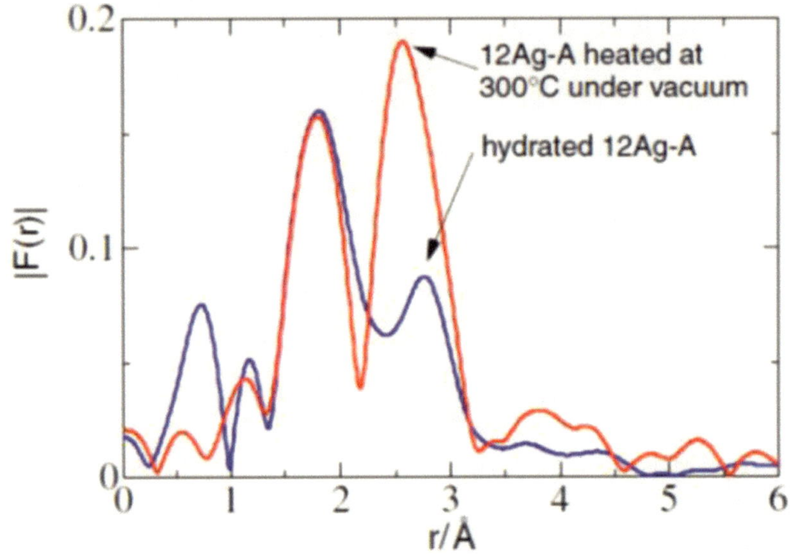

Figure 3. Fourier transforms (FT) of the Ag *K*-edge EXAFS spectra for the hydrated Ag-zeolite A measured at RT atmosphere and measured at RT after heating at 300 °C under vacuum [11].

Figure 4 shows $k^2\chi(k)$ spectra measured by quick EXAFS mode in the process of heating to 500°C atmosphere. Each spectrum was measured about 50 s and the corresponding sample temperatures are presented. The large structural change appears at around 150–300°C.

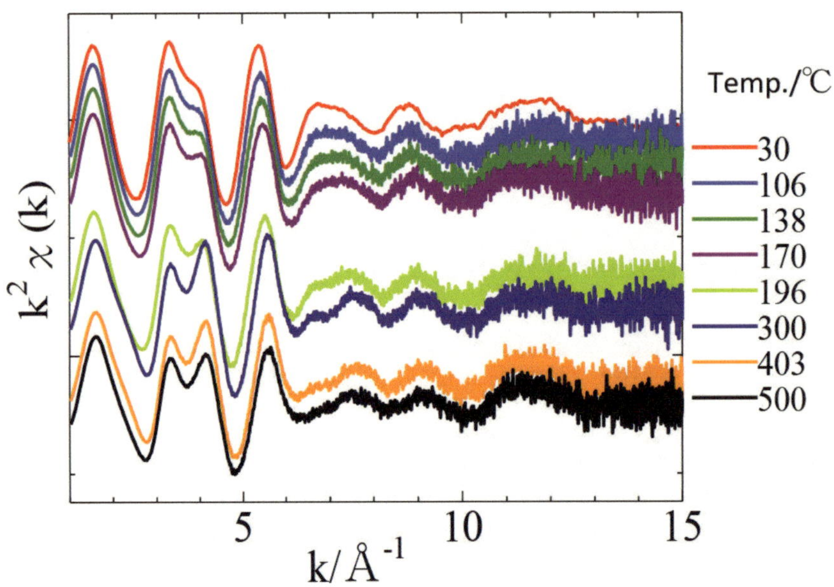

Figure 4. $k^2\chi(k)$ spectra for quick EXAFS measurement in the process of heating to 500°C atmosphere.

http://dx.doi.org/10.5772/63904

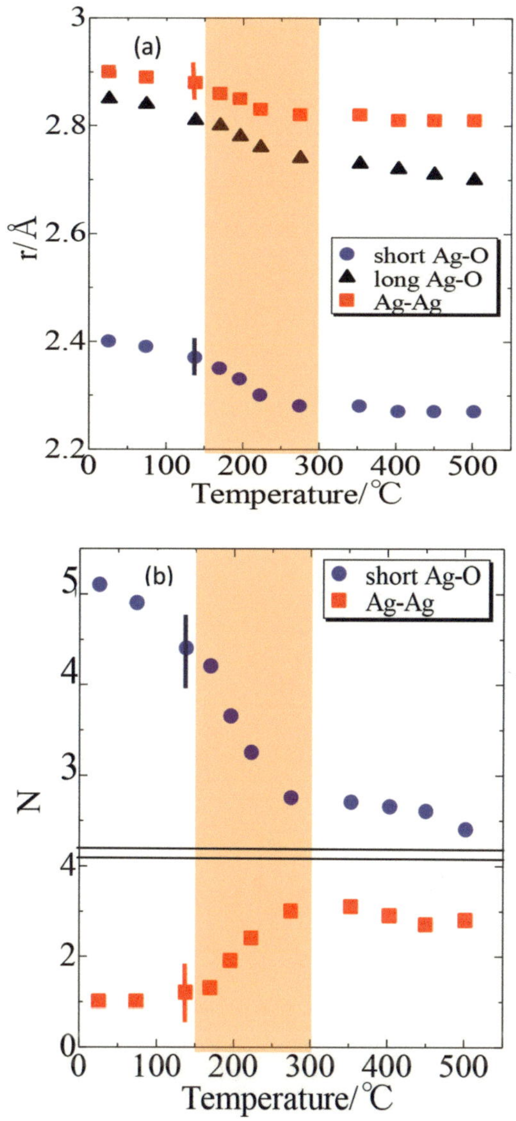

Figure 5. (a) Plots of the interatomic distances, *r*, and (b) that of the coordination number, *N*, as a function of measurement temperature.

The structural parameters obtained from the curve-fitting method by the use of Eq. (2) are plotted in **Figure 5(a)** for the interatomic distance, *r*, and **(b)** for the coordination number, *N*.

The Ag clustering in zeolite A occurs in the temperature range from 150 to 300°C. In this temperature range, the distance of Ag-O (short) decreases because of dehydration, and that of Ag-Ag also decreases because of less repulsion force between Ag^+ and Ag^0 when Ag^+ reduced by the dehydration. At the same temperature range, the increase of *N* for Ag-Ag indicates the Ag clustering process.

5. Mechanism of photoluminescence from Ag-zeolite A composites

In this section, we present the observation of PL and discuss the origin of the luminescent point in the Ag zeolite by the use of XAFS and IR. Many researchers believe that the luminescent points are Ag clusters themselves in the zeolite, but our result suggests that the origin of luminescence is the zeolite lattice after affected by the formation of Ag clusters in it.

First, we discuss the result of the PL measurement. **Figure 6** shows the PL curves for various introduction gases after heating of Ag-zeolite A in vacuum. The wavelength of the excitation light is 405 nm. The strongest intensity of PL band around 2.1 eV was observed when the air or the mixture of the $H_2O + N_2$ is introduced. It is suggested that the combination of $H_2O + N_2$ is important for the strong PL band. On the other hand, PL is not observed when the O_2 was introduced. In the present stage, the role of the gases of O_2, N_2, and H_2O for these phenomena has been unclear.

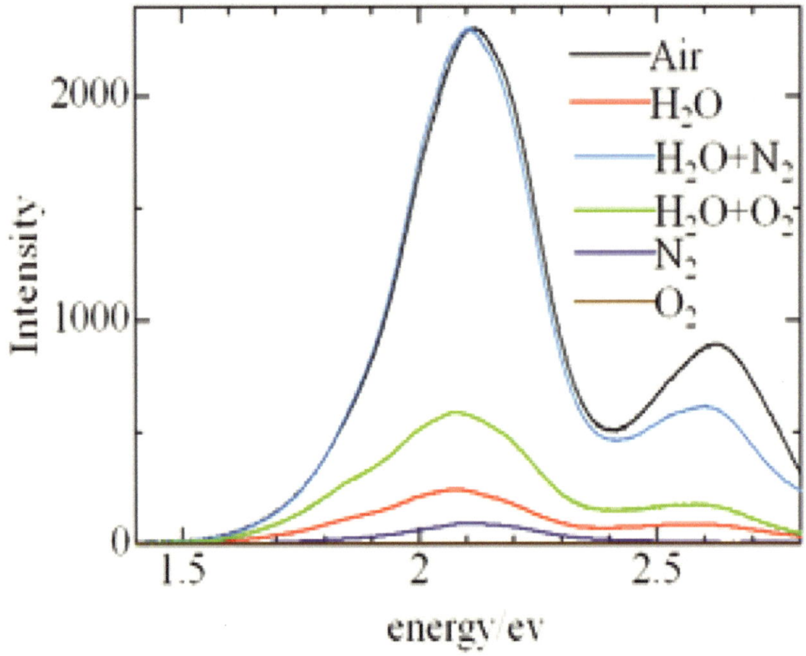

Figure 6. PL curves for various introduction gases after heating of Ag-zeolite A. The wave length of the excitation light is 405 nm [12].

Figure 7 shows the EXAFS $k^2\chi(k)$ spectra for Ag-zeolite A after following gases were introduced: Air, O_2, N_2, $H_2O + O_2$, $H_2O + N_2$ with unheated state. The structure around 3.5 Å^{-1} is returned to unheated state when air or H_2O+N_2 was introduced. This reversible change in $k^2\chi(k)$ spectra indicates the breakdown of the Ag clusters. On the other hand, the $k^2\chi(k)$ spectra were not changed when O_2 gas was introduced.

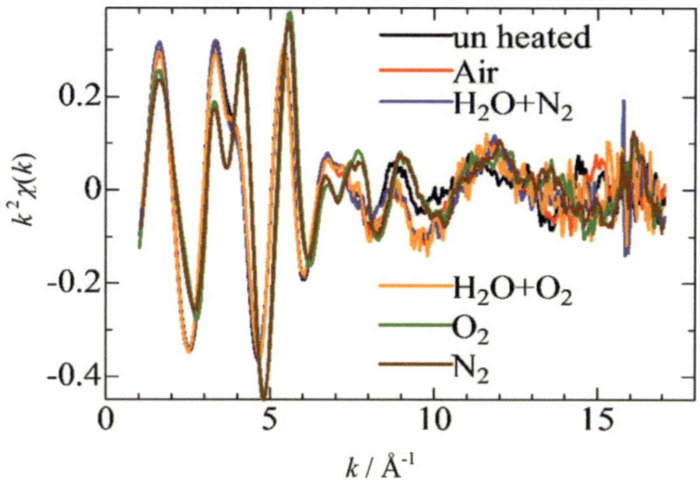

Figure 7. $k^2\chi(k)$ spectra for Ag-type zeolite A after the following gases were introduced: Air, O_2, N_2, $H_2O + O_2$, $H_2O + N_2$ with unheated state.

Figure 8 shows the Ag *K*-edge Fourier transform spectrum for Ag-zeolite A measured at room temperature in atmosphere and that after following gases were introduced: Air, O_2, N_2, $H_2O + O_2$, $H_2O + N_2$ with unheated state. The peak at ca 1.7 Å is assigned to the first nearest neighbor (1NN) O1 atom, and the second peak at ca 2.7 Å is to the second nearest neighbor (2NN) O2 and Ag atom. The three-shell curve-fitting technique was applied and the unknown structural parameters, the atomic distance *r* and the coordination number *N* for Ag-O1, Ag-O2, and Ag-Ag, were determined.

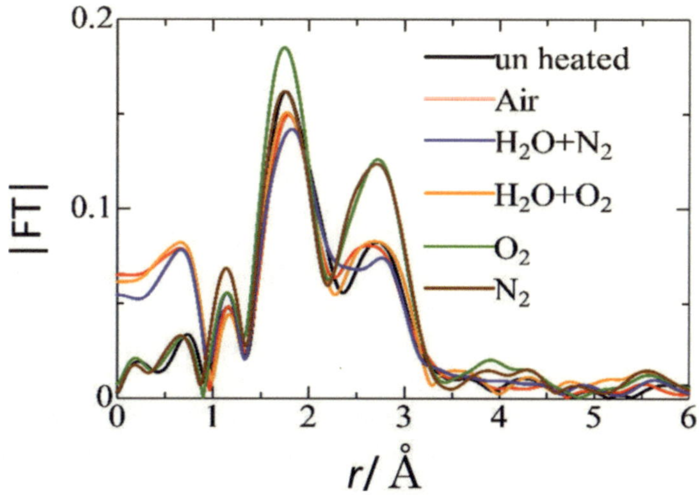

Figure 8. Ag *K*-edge Fourier transform spectrum for Ag-zeolite A measured at room temperature in atmosphere and that after following gases were introduced: Air, O_2, N_2, $H_2O + O_2$, $H_2O + N_2$ with unheated state [12].

Table 1 shows the structural parameters for Ag-zeolite A after the various gases are introduced. We can see in **Figure 6** that after the Air is introduced the strong PL band is observed. The Ag clusters were broken down according to the introduction of the Air. The parameters of r_{Ag} and N_{Ag} are very close to those in atmospheric pressure for the room temperature, which mean no Ag clusters. It takes enough time as 24 hours to breakdown Ag cluster by the introduction of Air. The present PL and XAFS results lead to the fact that the existence of the Ag clusters inhibits the strong PL band. When the O_2 gas is introduced, no PL band is observed. The r_{O1} and r_{Ag} are constant after the O_2 gas is introduced, and the N_{Ag} is still large (N_{Ag} ~2.35), which indicates that the Ag clusters still exist. We consider the cluster change after the mixed gases of H_2O+O_2, and H_2O+N_2 are introduced. When H_2O+O_2 mixture gases were introduced, weak PL band is detected. On the other hand, strong PL band is observed in the case of $H_2O + N_2$. The $N_{Ag\text{-}Ag}$ evaluated from the curve-fitting method shows the breakdown of the Ag clusters similar to the case of the air or both cases $H_2O + O_2$ and $H_2O + N_2$ introduction.

	r_{O1}(Å)	N_{O1}	r_{O2}(Å)	N_{O2}	r_{Ag}(Å)	N_{Ag}
Air	2.38	4.57	2.86	2.96	2.86	1.40
O_2	2.30	4.14	2.85	4.85	2.84	2.35
$H_2O + O_2$	2.39	6.73	2.86	5.19	2.88	0.94
$H_2O + N_2$	2.39	5.75	2.86	4.23	2.87	1.23

Table 1. The structural parameters, *r* and *N* after the various gases are introduced [12].

Figure 9 shows the IR result of Ag-zeolite A. The black line shows the IR transmission spectrum of the unheated Ag zeolite. Since the zeolite has a complicated structure and bands are superimposed on the various Si-O and Al-O, it is extremely difficult to distinguish between them. Therefore, assignment of the absorption band around 1000 cm^{-1} region is only represented as ν(T-O) (T = Si or Al) [16]. Bands appearing at relatively high wavenumber are assigned to the $\nu_{a\sigma}$ mode, and the low wavenumber region is assigned to ν_{σ} mode. Bending mode appears in the lower-energy region, not shown here.

The green line shows that the spectrum of the Ag zeolite was cooled to RT after heating for 24 h at 500 °C in vacuum (measured in a vacuum). As compared to the case of unheated, absorption band has been changed drastically. While individual band assignment is not known as mentioned above, considering the results of XAFS, it is reasonable to assume the change due to the formation of Ag clusters. For cluster formation, it is necessary that Ag^+ ions (at least partially) are reduced to Ag^0; also Ag which is to be aggregated away from the site of Ag^+ ion was located. Reduction of ions and changing the position also affect vibration of the framework in each case. (The distance changes between the framework and the Ag atoms and changes in the electrostatic attraction of Ag^+ can perturb the vibration of the T-O. In other words, it can be said that the IR is a sensitive technique to such a change.)

The red line is the spectrum in the case of introducing Air into the zeolite, cooling to RT after the vacuum heating. The spectrum greatly changed again, and is much closer to the spectrum of the unheated spectrum indicated in the black line. While XAFS for the state that Air intake after heating substantially coincide with that unheated, IR spectra for both state do not mach completely each other. Distinct band appeared at 1140 cm^{-1} and the strength of the main band of 1000–900 cm^{-1} decreased slightly. The cause of this change is not made apparent. However, the state where PL is expressed, that is, in the state where the excitation light is illuminated, the strength of the main band increases and 1140 cm^{-1} band decreases (**Figure 10**). For these reasons, this band change is evident and directly related to PL. As a cause of this change, considering that a change in XAFS is not almost observed, there may be two hypotheses as follows:

(1) The oxygen deficiency generated in the framework by treatment such as heated in vacuum.

(2) After Ag clusters collapse, Ag^+ cannot be located in a stable site, "caught" in a metastable site located in the vicinity of stable site. (This metastable site needs a shift of the slight position that cannot be detected by XAFS.)

Since Ag loading into zeolite performed in solution for a sufficient time, Ag^+ ions can be positioned at a stable site. By contrast, disintegration after cluster formation is in atmosphere at room temperature and may not exceed a potential barrier existing between the stable site and the metastable sites.

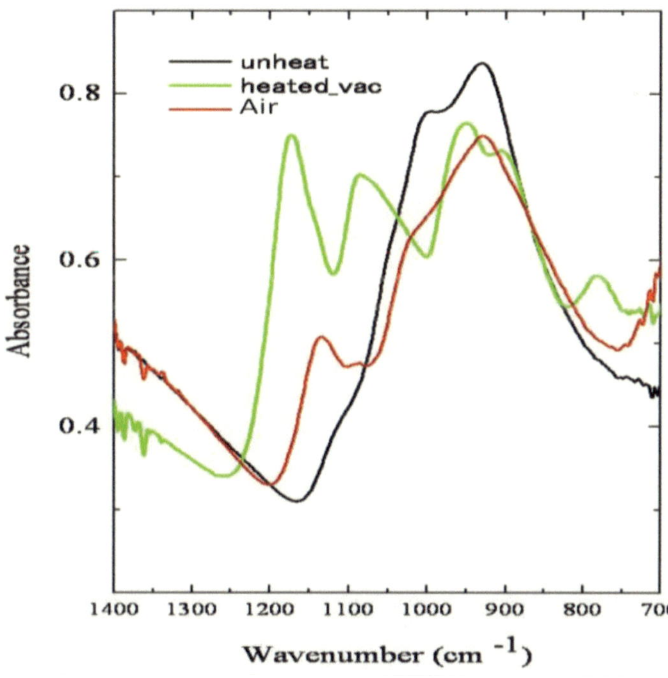

Figure 9. IR spectra for Ag-zeolite; unheated, after heating to 500 °C in vacuum and air introduced after heating.

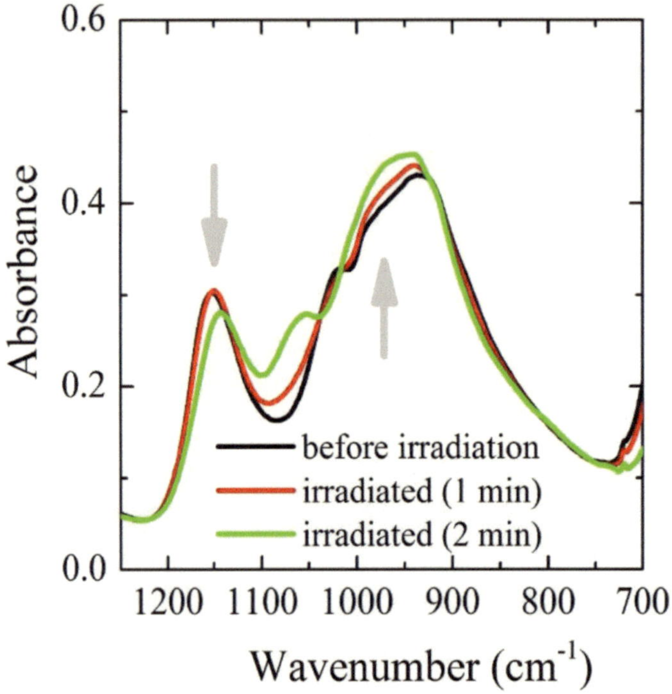

Figure 10. IR spectra for Ag-zeolite; Air introduced after heating (before irradiation), irradiated by excited light (405 nm, 1 min) and irradiated by excited light (405 nm, 2 min).

6. Summary

In this chapter, we discussed the formation of Ag clusters in the Ag-zeolite A cavity and the origin of strong PL after the formation of Ag clusters in zeolite A.

Ag clusters are formed in the cavity of Ag-substituted zeolite A after heat treatment under atmosphere or in vacuum. In the cavity, Ag^+ is reduced by the removal of the water molecule and Ag-O distance decreases. The Ag-Ag distance is also decreased because the repulsion between Ag and Ag is weakened. The coordination number of Ag-Ag is increased to around 2.35 after producing Ag clusters.

On the other hand, strong PL appears at 2.1 eV excited by 405-nm laser with Air or mixed gas of H_2O and N_2. The present Ag *K*-edge XAFS study indicated the local structure of Ag atoms in chemical species, which shows that the strong PL is almost the same as unheated one and Ag clusters are broken down. The IR spectra in the region of zeolite framework vibration of the species after Ag cluster breakdown are different from the unheated one. This indicates that the framework structure of zeolite changed by the production and breakdown of Ag clusters in the cavity. The effect to zeolite framework has a key for the strong PL.

http://dx.doi.org/10.5772/63904

Acknowledgements

The authors are grateful to Prof. H. Hoshino for the fruitful discussion of XAFS and PL in Ag-zeolite A. The synchrotron radiation experiments were performed at the Photon Factory in KEK.

Author details

Takafumi Miyanaga*, Yushi Suzuki and Sho Narita

*Address all correspondence to: takaf@hirosaki-u.ac.jp

Department of Mathematics and Physics, Hirosaki University, Aomori, Japan

References

[1] Y. Kim, K. Seff. The octohedral hexasilver molecule. Seven crystal structures of variously vacuum-dehydrated fully Ag^{+}-exchanged zeolite A. J. Am. Chem. Soc. 1974;100:6989.

[2] T. Sun, K. Seff. Silver clusters and chemistry in zeolites. Chem. Rev. 1994;94(4):857.

[3] R. Seifert, R. Rytz, G. Calzaferri. Colors of Ag^{+}-exchanged zeolite A. J. Phys. Chem. A. 2000;104:7473.

[4] G. Calzaferri, C. Leiggener, S. Glaus, D. Schurch, K. Kuge. The electronic structure of Cu^{+}, Ag^{+}, and Au^{+} zeolites. Chem. Soc. Rev. 2003;32:29.

[5] G. De Cremer, E. C-Gonzalez, M.B.J. Roeffaers, D.E.De Vos, J. Hofkens, T. Vosch, B. F. Sels. In situ observation of the emission characteristics of zeolite-hosted silver species during heat treatment. ChemPhysChem. 2010;11:1627.

[6] E. C-Gonzalez, M.B.J. Roeffaers, B. Dieu, G. De Cremer, S. Leyre, P. Hanselaer, W. Fyen, B. Sels, J. Hofkens. Determination and optimization of the luminescence external quantum efficiency of silver-cluster zeolite composites. J. Phys. Chem. C. 2013;117:6998.

[7] G. De Cremer, E. C-Gonzalez, M.B.J. Roeffaers, B. Moens, J. Ollevier, M. Van der Auweraer, R. Schooheydt, P.A. Jacobs, F.C. De Schryver, J. Hofkens, D.E. De Vos, D.F. Sels, T. Vosch. Characterization of fluorescence in heat-treated silver-exchanged zeolites. J. Am. Chem. Soc. 2009;131:3049.

[8] T. Miyanaga, H. Hoshino, H. Endo, H. Sakane. EXAFS study of silver clusters in zeolite. J. Synchrotron Rad. 1999;6:442.

[9] T. Miyanaga, H. Hoshino, H. Endo. Local structure of silver clusters in the channels of zeolite 4A. J. Synchrotron Rad. 2001;8:557.

[10] Y. Suzuki, T. Miyanaga, H. Hoshino, N. Matsumoto, T. Ainai. In-situ XAFS study of Ag clusters in zeolite 4A. Phys Scr. 2005;T115:765.

[11] H. Hoshino, Y. Sannohe, Y. Suzuki, T. Azuhata, T. Miyanaga, K. Yaginuma, M. Itoh, T. Shigeno, Y. Osawa, Y. Kimura. Photoluminescence of the dehydrated Ag-type zeolite A packed under air. J. Phys. Soc. Jpn. 2008;77(6):064712.

[12] A. Nakamura, M. Narita, S. Narita, Y. Suzuki, T. Miyanaga. In-situ XAFS study of Ag clusters in Ag-type zeolite-A. J. Phys. Conf. Ser. 2014;502:012033.

[13] H. Sakane, T. Miyanaga, I. Watanabe, N. Matsubayashi, S. Ikeda, Y. Yokoyama. Reproducibility tests of extended X-ray absorption fine structure for aqua and amine complexes of first transition metals in solid and aqueous solution. Jpn. J. Appl. Phys. 1993;32:4641.

[14] Available from: http://cars.uchicago.edu/ifeffit/Downloads 2016.5.21

[15] A.L. Ankudinov, B. Ravel, J.J. Rehr, S.D. Conradson. Real space multiple scattering calculation of XANES. Phys. Rev. B. 1998;58:7565.

[16] S.A. Rodrigues. Vibrational spectroscopy and structural analysis of Na–Y zeolite. Vib. Spectrosc. 1995;9:225.

Chapter 9

Zeolites as Potential Structures in Obtaining Jet Fuel Through the Fischer-Tropsch Synthesis

Lidiane Sabino da Silva,
Victor Luis dos Santos Teixeira da Silva,
Maria Angélica Simões Dornellas de Barros and
Pedro Augusto Arroyo

Additional information is available at the end of the chapter

http://dx.doi.org/10.5772/63662

Abstract

Due to the constant expansion of the aviation sector, the global air industry has performed the search for alternative fuels to petroleum-derived aviation kerosene, which present low emission of greenhouse gases and other pollutants, in addition to supplying the engines and aircraft already in operation using the existing distribution infrastructure. Thus, one alternative that has been studied is the synthetic kerosene derived from gasification processes followed by Fischer-Tropsch synthesis. Several features seem to be crucial in controlling product selectivity in the aviation fuel range, among them are the zeolitic support, acidity, and promoters. Therefore, the effect of these parameters in the Fischer-Tropsch synthesis is discussed in this chapter and, finally, the zeolitic catalysts that, according to the literature, have significant potential in obtaining synthetic aviation fuel are evidenced.

Keywords: zeolites, aviation kerosene, Fischer-Tropsch synthesis, acidity, promoters

1. Introduction

The air transport sector, due to constant expansion, is being pressured to reduce their CO_2 emissions, since aviation is responsible for 10% of the emissions in transport sector. In addition, because of the prospects for growth in this sector, there is evidence of a further increase in CO_2 emissions in the coming years [1]. For this reason, the airlines associated with the International Air Transport Association (IATA) pledged to improve fuel efficiency by 1.5% annually to 2020, cap net emissions from 2020 with carbon-neutral growth and cut net emissions in half by 2050 compared to 2005 [2]. Such obligations, in the future, may limit the civil aviation market and stimulate the search for alternatives. However, the demand for fuels tends to expand, since the passenger demand will be more than doubling from 3.3 to 7.3 billion by 2034 [3], even with the possible gains in energy efficiency (estimated between 30 and 50%), from improved aerodynamics, reducing the weight of the aircraft, the advances in technology of thrusters, and the adoption of best practices in air traffic management [4].

In fact, historically, since the first aircraft was manufactured, the airline industry has been trying to decrease the fuel consumption through the improvement of technologies used. Although there is a decrease in emissions over the years, due to new technologies employed, the speed of which can perform this benefit is compromised by the fact that airlines maintain their assets in use for about 40 years, due to the high investment cost. Thus, the new technologies have to wait until the fleet is replaced [5].

In this context, the global air industry has performed the search for alternative fuels to petroleum-derived aviation kerosene, which present low emission of greenhouse gases and other pollutants, in addition to supplying the engines and aircraft already in operation using the existing distribution infrastructure [6]. Therefore, one alternative that has been studied is the synthetic kerosene derived from gasification processes followed by Fischer-Tropsch synthesis (FTS). In this synthesis route, the raw material (coal, natural gas, or biomass) is transformed into syngas, a mixture of carbon monoxide and hydrogen, which is converted by means of a metal catalyst on FTS, forming hydrocarbons of different sizes. The final step is the processing and fractionation of products to produce a fuel with appropriate properties to be used in the airline industry [7].

The literature shows that the structure of the support used in the catalyst can be one of the key features in obtaining products in the aviation fuel range. When the catalytic reaction occurs within the pores of a zeolite, the size and shape of the canals and cavities can be used to select the desired reaction path, making use of "restricted transition state selectivity" or, also, through the "product selectivity." In the first case, the reaction intermediates and/or transition states are sterically limited due to the shape and size of the microporous lattice [8]. In the second case, the product selectivity refers to discrete diffusivities of the reaction products formed with respect to the microporous architecture and crystal size of the catalyst particles. If the molecules formed are very large compared to the pore size, the product diffusion may not occur, be extremely slow, or be completed after the occurrence of secondary reactions, such as cracking and isomerization, among others [8, 9]. In this way, specific structures with unique characteristics can direct the formation of products with particular chain length.

Another feature that seems to be crucial in controlling product selectivity is the acidity. The acid sites of the zeolites can promote cracking, isomerization and aromatization reactions. This property can help in overcoming the limitation of Anderson-Schulz-Flory (ASF) distribution and adjustment of the Fischer-Tropsch products in a specific distribution [10]. This can also reduce the need for subsequent treatments to Fischer-Tropsch synthesis, which could lead to a significant reduction in process costs [11].

Finally, one must take into consideration the use of specific promoters, which can provide desirable characteristics to the catalysts, such as increasing the reaction rate and greater stability and selectivity of the hydrocarbons C_{5+} [12, 13]. The promoters may direct the formation of products at a particular chain length, especially when combined with a structure and a specific acidity [11, 14].

However, despite the wide literature directing the production of synthetic fuels, there is still much to be studied on the aviation kerosene, since the vast majority of articles have focused on the production of gasoline and/or diesel [15–18], or simply the formation of products in the range of liquid hydrocarbons C_{5+} [13, 19–23]. For this reason, the main objective of this chapter is to analyze zeolitic structures studied in the literature, besides the influence of acidity and promoters, which have significant potential in obtaining fuel in the range of aviation kerosene (C8–C16) through the Fischer-Tropsch synthesis.

2. Conventional aviation fuel

The petroleum refining occurs through a process of fractional distillation, in which crude oil is heated and its different components come boiling at various temperatures, and later being condensed. Therefore, specific components are retrieved in different temperature ranges.

The lowest boiling fraction, taken from the top of the distillation column, is called naphtha, which is mainly processed further to make motor gasoline. The second fraction of about 33% of the crude oil input contains the raw material for jet fuel production. This fraction is further processed in the distillate hydrotreater to obtain the desired products. Basically, kerosene is originated as a straight-run (distilled) petroleum fraction with boiling temperature ranging from 205 to 260°C [24]. Subsequent fractions correspond to components in the range of diesel fuel and residues, and the latter can be distilled under vacuum, yielding high-value distillate components [25].

The petroleum processing through fractional distillation is still the dominant technology in the production of aviation kerosene; besides, this fuel accounts for 6.3% of the total amount of crude oil consumed [24]. However, the large fluctuations in the price of petroleum, the concern about the pollutants emitted, and the increase in passenger demand show a strong incentive for the airline industry to invest in alternative sources of fuel.

3. Synthetic aviation fuel

The airline industry bet on fuels that are chemically indistinguishable from conventional fuel and can be mixed in order to provide similar performance and security, without change in the aircraft systems and engines or on distribution and storage infrastructure. One of the most promising alternative processes for obtaining aviation kerosene is the Fischer-Tropsch process.

Currently, Semi-Synthetic Jet Fuel (SSJF) obtained by Fischer-Tropsch process and hydroprocessing, mixed with JET-A1, is already approved by ASTM D7655, limited to 50% proportion.

One of the reasons that the SSJF fuel cannot be used without restriction in civil aviation, especially those derived from natural gas or biomass, is the fact that these products do not present aromatic hydrocarbons, not answering, so, to the specifications imposed by the existing standard (minimum of 8% aromatics) [26].

The lack of aromatic components in the fuel results in higher freezing points and lower densities. The freeze point of a fuel composition is an important factor in determining whether it is suitable for aviation use, for which low-temperature conditions are experienced at high altitudes. Therefore, it is vital that the fuel composition does not freeze or cause flow to be restricted during operation; otherwise, the consequences could be disastrous [27]. In addition, the deficiency of aromatics in aviation kerosene can also cause shrinking of elastomers present in the storage and distribution system, which can lead to fuel leaks [28].

One of the advantages of using semi-synthetic kerosene is cleaner burning, since there is a lower emission of sulfur compounds [28], leading to an increase in the useful life of the combustor and turbine. Moreover, the emission of particulate compounds is lower compared to conventional fuel [24]. Therefore, the synthetic fuel blend with the conventional allows aggregating the advantages of synthetic fuel, while it is appropriate in relation to pre-established standards.

Currently, there is also the Fully Synthetic Jet Fuel (FSJF). The SASOL was the first and only one to produce this type of fuel for unrestricted use in civil aviation, from 2008, proving that the fuel produced behaved as the conventional in all aspects of storage, handling, and flight performance [7].

Coal-derived fuels are rich in naphthene compounds [29]. Due to the historic necessity by SASOL to produce diesel and gasoline fuel from their coal reserves located in South Africa, the catalysts and operating conditions have a different focus than other plants currently online and/or being developed [7]. This fact probably contributed to the success in obtaining the Fully Synthetic Jet Fuel from the Fischer-Tropsch synthesis.

The process consists of a mixture of five main blends, of which fractions of kerosene can be obtained. These blends come from the Fischer-Tropsch process or products generated in coal gasification. Therefore, the blend in greater amount is obtained from the Fischer-Tropsch process. The other four blends are obtained from heavy fractions of naphtha, commonly used in the production of gasoline, or light fractions of distillate, usually used in the production of diesel [7].

4. The Fischer-Tropsch synthesis

4.1. The Fischer-Tropsch reaction

The Fischer-Tropsch process has four main steps. The first stage corresponds to the production of synthesis gas, which is a mixture of hydrogen and carbon monoxide. In the second step, the removal of unwanted compounds and carbon dioxide from syngas occurs. The third step corresponds to the Fischer-Tropsch reaction, in which the formation of hydrocarbons occurs. After this step, the products undergo treatments of hydrocracking and isomerization in order to acquire the necessary features to be used as fuel.

Among the products obtained in Fischer-Tropsch synthesis are the alkenes (olefins), alkanes (paraffins), and oxygenates (e.g., alcohols, aldehydes, carboxylic acids, and ketones), in addition to water [30]. The nature of oxygenated compounds is determined by the catalyst type and reaction conditions. The paraffins are generally straight chains, forming products in the range of gasoline, diesel oil, aviation kerosene, and waxes [31].

The temperature and reaction pressure are some of the most important parameters in the SFT to obtain products at a specific desired range. Currently, there are some well-defined conditions: the Low-Temperature Fischer-Tropsch (LTFT), the High-Temperature Fischer-Tropsch (HTFT), and pressures in the range of 10–40 bar. In LTFT process, iron or cobalt catalysts are typically used at approximately 230°C, yielding mainly products in the kerosene range, diesel, and waxes. In the HTFT process, iron catalysts are typically used at approximately 340°C, yielding mainly products in the range of gasoline and olefins [32]. These different behaviors occur because at elevated temperatures the reaction rate and quantity of secondary products increases, which leads to a reduction in the average chain length of the molecules, with a corresponding increase in methane formation [33].

On the operating pressure, the literature shows that the increase in the value of this parameter provides an increase in carbon monoxide conversion and C_{5+} selectivity [11, 34–36]. The increased conversion of carbon monoxide is explained by the greater amount of C_1 intermediates, which increase the propagation rate [35]. In the same context, the increase in C_{5+} selectivity occurs due to the higher concentration of syngas, leading to a higher propagation rate of the carbon monoxide polymerization reaction and, therefore, higher concentrations of long-chain hydrocarbons [37].

The overall reaction of Fischer-Tropsch synthesis is highly exothermic and is subject to limitation diffusional, which can affect the distribution of final products and lead to the deactivation of catalysts [38]. The overall reaction is represented below:

$$nCO + 2nH_2O \rightarrow \left(CH_2\right)n + nH_2O \qquad \Delta = -167.4kJ/mol \tag{1}$$

Secondary reactions, according to the literature [39, 40], are represented as

Paraffins

$$nCO + \left(2n+1\right)H_2 \rightarrow C_nH_{2n+2} + nH_2O \tag{2}$$

Olefins

$$nCO + 2nH_2 \rightarrow C_nH_{2n} + nH_2O \tag{3}$$

Alcohols

$$nCO + 2nH_2 \rightarrow C_nH_{2n+1}OH + (n-1)H_2O \tag{4}$$

Methane

$$nCO + 3nH_2 \rightarrow CnH_{4n} + nH_2O \tag{5}$$

Water-gas shift reaction

$$nCO + nH_2O \rightarrow nCO_2 + nH_2 \tag{6}$$

Coke deposition

$$nCO + nH_2 \rightarrow nC + nH_2O \tag{7}$$

Boudouard reaction

$$nCO + nCO \rightarrow nC + nCO_2 \tag{8}$$

The distribution of products in the Fischer-Tropsch synthesis follows the Anderson-Schulz-Flory distribution [41], which is expressed as follows:

$$W_{n/n} = (1-\alpha)^2 . \alpha^{(n-1)} \tag{9}$$

where $W_{n/n}$ is the weight fraction of hydrocarbon molecules containing n carbon atoms, and α the chain growth probability or the probability of molecule to continue reacting to form a longer chain.

The product selectivity is then determined by an ideal α value. Smaller α values are obtained when short-chain hydrocarbons are produced (C_1–C_4), while higher values of α are obtained when heavier hydrocarbons are formed (C_{21+}). Therefore, the Anderson-Schulz-Flory model is not selective for the hydrocarbon's intermediate range, which is in the vast majority of the cases the most desired products [42]. For example, according to the Anderson-Schulz-Flory model, the maximum product selectivity in gasoline range (C_5–C_{11}) and diesel (C_{10}–C_{20}) is approximately 45 and 35%, respectively.

According to Zhang et al. [42], the catalytic factors that exert greater influence in the activity and selectivity of catalytic in the FTS are the type of support, the promoters, and the active phase and its size, chemical state, and microenvironment in which it inserts. Therefore, the development of selective catalysts, which can direct the formation of hydrocarbons in the desired range, is one of the biggest challenges in the synthesis of catalysts for Fischer-Tropsch synthesis.

4.2. Influence of zeolitic support

The support structure used in the catalyst can be one of the key features in obtaining products in a specific range. When the catalytic reaction occurs within the pores of zeolite, the size and shape of the channels can be used to select the desired reaction path making use of the "shape-selectivity." This occurs when the spatial configuration around reactants, products, transition state, or a reaction intermediate located in the intra-crystalline volume is such that only certain configurations are possible [8]. Therefore, specific supports may direct the formation of products with a particular chain length.

Molecular shape selective is observed frequently when using zeolite catalysts because of their two- or three-dimensional nature. Zeolites have unique advantages which render them particularly attractive for a variety of conversions and processes of industrial interest [9]:

- zeolites have pore dimensions comparable to molecular size and reactions occur within their intra-crystalline space;
- the active center concentration, exchanger cations or metals with catalytic properties and/or acidic sites, usually depends directly on their aluminum content, which can be varied;
- zeolites are thermally stable, which enables their operation as catalysts in a broad temperature range.

An advantage of the catalysts with shape selectivity is the formation of products not predicted in the equilibrium composition. The thermodynamic equilibrium laws are valid only if the catalytic and non-catalytic reactions are truly the same, and yield identical products. However, these considerations may not be obeyed in chemical processes. The catalytic reactions inside the pore channel of a shape-selective catalyst can be substantially different from those that occur in homogeneous catalytic or non-catalytic reactions involving the same reactants, due to the selectivity mechanisms characteristic of each structure [43].

The selectivity mechanisms of catalysts with shape selectivity are described briefly below:

- Reactant selectivity: it describes the phenomenon in which microporous catalysts act as molecular sieves and exclude molecules larger than the opening of pores in the structure, while allowing the entry of smaller molecules; this type of selectivity depends on intra-pore diffusional characteristics of reacting molecules [43];
- Product selectivity: it refers to the selective formation of certain products, when there are other potential products whose formation is also thermodynamically feasible but is limited because of their limited diffusion out of pore; smaller molecules can diffuse through the micropores, while larger molecules require more time to remain inside the structure [43];

this allows these molecules to undergo secondary reactions of isomerization and cracking, among others;

- Restricted transition state-type selectivity: it occurs when the spatial configuration around a transition state or a reaction intermediate located in the intra-crystalline volume is such that only certain configurations are possible; this means that the formation of reaction intermediates and/or transition states is sterically limited due to the shape and size of the microporous lattice allowing the access of the species formed to interact with the active sites [8].

Despite the vast majority of articles presenting the formation of gasoline or diesel fuel through the Fischer-Tropsch synthesis as an objective, some zeolites can be identified with considerable potential in the production of hydrocarbons in the aviation kerosene range. This stems from the fact that these structures present structural and acidic properties that give significant shape selectivity to products or transition states to the catalysts.

Bessel [44] investigated bifunctional catalysts of cobalt supported in zeolites ZSM-5, ZSM-11, ZSM-12, and ZSM-34 to the Fischer-Tropsch synthesis. The use of bifunctional catalysts composed of active metal and zeolite in acidic form aimed to overcome the typical nonselective Anderson-Schulz-Flory product distribution arising from conventional catalysts. In this type of catalyst, the active metal is responsible for the growth of the primary straight-chained hydrocarbon product, which undergoes further restructuring at the zeolite acid sites to produce a more branched hydrocarbon product of limited chain length. Despite having the lowest conversion of carbon monoxide (45%), among the catalysts studied, Co/ZSM-34 showed the highest selectivity to hydrocarbons in the jet fuel range (over 30%). This result was not expected because the ZSM-34 was the zeolite that showed the highest acidity; however, this also presents the more complex structure, with the largest (6.7 Å) and smallest (3.6 Å) diameters of channels among the zeolites used. In this case, the accessibility to the active sites and pore structure appears to be more important than the acidity in obtaining products in the aviation kerosene range.

Zola et al. [45] synthesized cobalt catalysts with 10% metallic phase supported through wet impregnation in zeolites Beta, USY, mordenita, and ZSM-5. The catalyst supported on zeolite USY presented the highest carbon monoxide conversion values (19%) and activity in Fischer-Tropsch synthesis. Hydrocarbon production in the jet fuel range also appears to be greater for this catalyst. The USY zeolite has a three-dimensional system of channels and large micropores, which favors the accessibility to catalytic sites. Therefore, these characteristics seem to favor the formation of products with a specific chain length in Fischer-Tropsch synthesis.

Ngamcharussrivichai et al. [17] investigated the performance of cobalt-based catalysts supported on silica, alumina, montmorillonite, and three zeolites (ZSM-5, USY, and MCM-22) in Fischer-Tropsch synthesis, 10 bar, and 230°C. The catalysts supported on USY and montmorillonite showed the carbon number distribution to be narrower and lower occurrence of waxes compared to catalysts supported on silica and alumina, because the zeolitic structure limits the growth of hydrocarbon chain. The material acidity also contributed to narrow the carbon number distribution; however, the zeolitic structure was still the most relevant feature.

Among the Co/SiO_2, Co/USY, Co/ZSM-5, and Co/MCM-22 catalysts, the second and the third ones showed the distribution of carbon number more similar to the aviation kerosene (C_8–C_{16}). In addition, when a support with acidic characteristics was used, the formation of C_{12+} significantly decreased, probably due to the higher occurrence of cracking reactions.

Martínez et al. [18] studied the catalytic properties of ZSM-5, MCM-22, IM-5, and ITQ-2 zeolites, all with 10-membered ring pore openings (10-MR) and Si/Al ratio similar. The study aimed to form products in the gasoline range through the Fischer-Tropsch synthesis. Hybrid catalysts (zeolite + silica) were tested for at least 7–8 h of reaction showing conversions around 50%. The presence of zeolites in hybrid catalyst increases the selectivity to products C_2–C_4 and gasoline (C_5–C_{12}), as well as reduces the selectivity to diesel (C_{13}–C_{22}) and waxes (C_{23+}). The hybrid catalysts with zeolites MCM-22 and IM-5 showed very similar products and the lower yields to iso-C_{5-8}. The decreased formation of C_{23+} and the greater and lesser C_{13-22} formation and C_{5-12}, respectively, among the hybrid catalysts, suggest that the MCM-22 and IM-5 zeolites may have a narrower distribution of products, which can provide a higher yield of hydrocarbons in the aviation kerosene range (C_8–C_{16}). These properties are probably related to the unique features of the zeolitic structures used, as, for example, the presence of supercages in the case of MCM-22.

Kibby et al. [46] performed the synthesis of a hybrid catalyst, which was composed of Ru and Co, as active metals, ZSM-12 and ZSM-5 zeolites, as supports, and alumina as base. The intention was to synthesize a catalyst which maximizes the formation of C_{5+} products, were free of the solid phase (waxes), and minimize the formation of gaseous products (C_1–C_4) in the Fischer-Tropsch synthesis, so the product would not need to undergo further hydrocracking process. The hybrid catalyst using ZSM-5 produced a greater quantity of gaseous products and the average molar mass of hydrocarbons was lower when compared to hybrid catalyst with ZSM-12. In addition, the catalyst using ZSM-12 yielded a larger amount of products in the aviation kerosene range (C_8–C_{16}) than the hybrid catalyst with ZSM-5. Apparently, ZSM-5 is more effective in the isomerization of C_4–C_9 olefins, while ZSM-12 is more active in isomerization of C_{10} olefins, the latter leading to the formation of products with a longer length of chain. The reasons for this behavior are still not very well known, although the straight channels of ZSM-12 can carry heavier olefins faster. The authors concluded that the understanding of metal deposition rules on the support, the morphology and structure of zeolites, optimization of acid functions, activation, and process conditions are essential for obtaining the desired hydrocarbon range during the Fischer-Tropsch synthesis.

4.3. Influence of acidity

The microporous structure of zeolites provides a large internal surface area that allows a large concentration of active sites accessible to reactants, together with notable selectivity effects related both to diffusion of reactants and products inside the pore system and to steric constraints on intermediates and transition states. All these features influence the overall catalytic performance of acid zeolites, and make difficult to establish an acidity scale and a clear correlation between acidity and catalytic activity [47].

Two types of acid sites, Brönsted and Lewis, can be formed on the zeolite surface. Thus, Brönsted acid sites are generated when Si^{4+} is isomorphically replaced by a trivalent metal cation such as Al^{3+}. This substitution creates a negative charge in the lattice that can be compensated by a proton [48]. These protons can be present in the medium depending on the reactants that were used in the material synthesis, exerting the function of compensation cations. To replace these cations and consequently form a Brönsted acid site, ion exchange procedure is conducted with an aqueous solution of ammonium chloride, followed by washing to remove residual chloride, and calcination [49]. At high temperatures, the remaining protons can acquire mobility and be lost as water molecules, followed by the formation of Lewis acid sites, due to the isomorphic substitution of Al by Si [8].

It is evident that the amount of acid sites in the zeolite is related to the Si/Al ratio. The smaller the Si/Al ratio, more aluminum is present in the structure and increases the amount of compensation cations and the number of acid sites when the zeolite is in protonic form. However, the generation of acidic sites is associated with the imbalance of charges generated by isomorphic substitution of different load cations. Therefore, the greater the number of aluminum atoms, more unbalanced will be the zeolitic framework [49]. Due to this fact, the reduction of Si/Al ratio makes the zeolite structure to become less stable and may even damage the crystalline arrangement, besides having influence on the strength of acid sites.

In fact, not all acid sites have the same acid strength, because this feature depends on the number of aluminum atoms near the aluminum atom, which supports the acid site [50]. The more isolated is the trivalent atom, the greater the acid strength.

From the change of the Si/Al ratio of the zeolitic structure, through either synthesis or post-synthesis treatments, it is possible to change not only the total number but also the electronic density on the bridging hydroxyl group, and consequently to change the acid strength of the Brönsted acid site [51]. The control of acid strength as well as the density of acid sites of zeolite catalysts has also led to successful catalysts and processes in the field of oil refining and petrochemistry [48].

For the incorporation and/or suitability of the acidic species of Lewis or Brönsted, there are many synthetic strategies that can be developed, such as impregnation, deposition, precipitation, calcination, and reduction, among others. Structural variations are diverse and numerous chemical species can be incorporated into the support, in composition, oxidation states, and chemical environments in different ways [49].

In order to evaluate the influence of acidity on catalysts, Bessel et al. [52] showed that acid zeolitic structures have a tendency to catalyze secondary reactions in the Fischer-Tropsch synthesis, from olefinic products. Secondary reactions include oligomerization, isomerization, cracking, and hydrogen transfer, thus obtaining more branched products with a shorter chain length compared to the catalysts using supports with less or no acidity.

Ravishankar et al. [53] synthesized cobalt catalysts using MCM-22 as support, through wet impregnation. The Fischer-Tropsch synthesis was performed at 280°C, 12.5 bar, and H_2/CO = 2. Samples with lower Si/Al ratio showed the most selectivity to methane. These results also confirm the higher cracking of the products and, consequently, a lower selectivity to C_{5+} when using catalysts that are more acidic. It is well known that Fischer-Tropsch synthesis selectivity depends on the conversion level and one would expect an increase of C_{5+} selectivity with increasing the carbon monoxide conversion. However, the C_{5+} selectivity values obtained for the cobalt catalyst supported on MCM-22 with high Si/Al ratio (200) and different cobalt loadings seem to be independent of the conversion level. The authors justify the increased C_{5+} selectivity to the particular properties of the support structure related to shape selectivity, active metal, and Si/Al ratio.

Therefore, it is possible to provide, in most cases, the occurrence of secondary reactions through the presence of acid sites, which may be desirable to obtain hydrocarbons in the aviation kerosene range. The amount and type of sites will depend on intrinsic characteristics of the support, the reaction conditions, and the metallic phase.

4.4. Influence of promoters

In addition to a support with adequate characteristics and good Si/Al ratio, the use of specific promoters may offer desirable features to catalysts, such as, for example, increased reaction rate, greater stability, and C_{5+} selectivity. According to Iglesia [54], promoters can be classified as chemical and structural, increasing the amount of active sites and the activity per site, respectively. The general effect has been postulated to be due to many factors, including (1) intimate electronic contact changing the local band structure of the metal, (2) ensemble-type geometric effects, (3) reducing deactivation by carbonaceous deposits, and (4) enabling more surface sites to be reduced by hydrogen spillover during the initial activation [55].

According to the literature, one of the promoters who presented good results when incorporated into cobalt catalysts was manganese. Dinse et al. [12] investigated the use of this promoter for cobalt catalysts in the Fischer-Tropsch synthesis. The catalytic performance of the samples was carried out at 220°C, $H_2/CO = 2$, and operating pressure between 1 and 10 bar. At low pressures, the addition of promoter in the catalyst led to a decrease in methane selectivity and increased C_{5+} selectivity, while the formation of C_2–C_4 hydrocarbons suffered no influence. At high pressures, in addition to a decrease in methane selectivity and increased C_{5+} selectivity, there was also an increase in C_2–C_4 selectivity, as well as a large increase in olefins/paraffins in C_2–C_4 range.

Dinse et al. [12] attributed the selectivities obtained for Co/SiO_2 and $Mn\text{-}Co/SiO_2$ catalysts to the equilibrium constant for CO adsorption, which is greater than the equilibrium constant for H_2 adsorption in the presence of manganese, resulting in a reduction in H_2/CO ratio and a consequent increase in chain length of products. Selectivity differences in different pressures are also assigned mostly to the H_2/CO ratio, which decreases with the increase of the total pressure of reaction and, also, to the increase of α-olefins reincorporation in chain growth process. Therefore, the manganese exerts significant influence in obtaining long-chain hydrocarbons, and again the importance of reaction conditions in obtaining hydrocarbons in a specific range is evident.

Khobragade et al. [13] synthesized CO/SiO_2 catalysts using K and CeO_2 as promoters, from the sol-gel method, in order to evaluate the performance in Fischer-Tropsch synthesis. Among the samples tested, the catalyst promoted with cerium presented the highest C_{5+} selectivity, while the catalyst promoted with potassium presented higher methane selectivity. The behavior of the catalyst promoted with potassium was expected by its use to be more suitable when using iron as active metal, due to the higher reaction temperatures required for this catalyst [56]. In addition, potassium can cause poisoning of catalysts using cobalt as active phase [20]. According to Khobragade et al. [13], the behavior of the catalyst promoted with CeO_2 occurs due to a decrease in the hydrogenation rate during the Fischer-Tropsch synthesis, in which the adsorption of H_2 is suppressed in favor of chain growth. This can occur due to partial reduction of CeO_2, which would change the electronic properties of cobalt atoms through oxidation. It was also verified that most of the products were in the diesel range, probably due to the increase of active carbon species ($-CH_2-$) in the catalyst. Cerium probably decreases the surface of metallic cobalt, resulting in a different chain growth and more favorable to the formation of C_{5+} hydrocarbons. This effect of the promoter is obtained because of the close interaction with the species of cobalt.

A way of aggregating the advantages of various promoters and brackets in the same catalyst is the use of hybrids and multifunctional catalysts. Teiseh et al. [14] showed that the use of a multifunctional hybrid catalyst ($Co-SiO_2/Mo-Pd-Pt-HZSM-5$) produces more hydrocarbons in kerosene range than that of a conventional catalyst (Co/SiO_2). In this case, the presence of molybdenum in the hybrid catalyst leads to a combination of carbonium ions (oligomerization of olefins), which contributes to the formation of paraffin in the kerosene range.

In this same context, Yan et al. [11] studied the performance of multifunctional catalysts in the Fischer-Tropsch synthesis, using alumina as support, cobalt and iron as active metals, and potassium and molybdenum as promoters ($K-Fe-Co-Mo/\gamma-Al_2O_3$). In this case, the potassium can increase the selectivity and stability of the catalysts in the presence of iron. The multifunctional catalyst was used with the intention to reduce the number of steps necessary to obtain a jet fuel by Fischer-Tropsch synthesis. The syngas used was obtained from biomass, which typically has a low H_2/CO ratio [57]. In this case, the simultaneous use of Fe and Co is advantageous because, due to the iron having activity in the water-gas shift reaction, there is a setting in H_2/CO ratio for higher values, in which the cobalt is more active.

After obtaining the products, distillation was conducted for the removal of short-chain hydrocarbons and waxes. The samples were collected on boiling-point range between 110 and 310°C. The products obtained after distillation step presented features quite similar to the conventional fuel, in relation to both the distribution of organic functions and the number of carbons. Furthermore, the properties of the distilled fraction have become quite similar to jet fuel.

5. Potential zeolitic catalysts in obtaining synthetic aviation kerosene

Table 1 presents the catalysts with significant potential in the production of jet fuel through the Fischer-Tropsch synthesis. The Si/Al ratio of zeolites, the temperatures and pressure reaction, and the H_2/CO ratio of the syngas and the carbon monoxide conversion are presented in the table.

References	Zeolitic catalysts	Si/Al molar ratio	Reaction conditions			CO conversion (%)
			Temperature (°C)	Pressure (bar)	H_2/CO ratio	
Bessel et al. [44]	Co/ZSM-34	>10	240	20	2	45
Ravishankar et al. [53]	Co/MCM-22	220	280	12.5	2	4.5
Zola et al. [45]	Co/USY	7	240	10	2	19
Ngamcharussrivichai et al. [17]	Co/USY	16	230	10	2	38
Ngamcharussrivichai et al. [17]	Co/ZSM-5	19	230	10	2	19
Martínez et al. [18]	Co-SiO_2/MCM-22	15	250	20	2	50
Martínez et al. [18]	Co-SiO_2/IM-5	14	250	20	2	50
Teiseh et al. [14]	Co-SiO_2/Mo-Pd-Pt-ZSM-5	23	250	10	1.9	88
Kibby et al. [46]	Co-Ru/ZSM-12/Al_2O_3	45	220	20	2	60

Table 1. Catalysts with significant potential in the production of jet fuel through the Fischer-Tropsch synthesis.

According to a review carried out, the zeolitic supports used in catalysts for Fischer-Tropsch synthesis wield great influence on the product's chain length. This occurs due to the fact that the support structure can inhibit or head the formation of certain products. In this case, the zeolite ZSM-34, USY, ZSM-5, ZSM-12, MCM-22, and IM-5 have significant potential in obtaining products in the aviation kerosene range. In addition, the Si/Al ratio in the catalyst, especially in the range of 7–220, involves the presence of acid sites, which may promote secondary reactions that reduce the formation of waxes and narrow the product's selectivity. Another factor that should be considered is the use of promoters, which in mass concentrations of about 1% may confer upon a major activity to the catalyst and assist in the formation of long-chain hydrocarbons. In addition, the Fischer-Tropsch process at low temperatures and pressures between 10 and 20 bar shows more appropriate in obtaining hydrocarbons C_8–C_{16}. Thus, the use of supports with appropriate structural characteristics, specific Si/Al ratio, specific promoters, and suitable conditions of operation shows promise to obtain products with features similar to the jet fuel through the Fischer-Tropsch synthesis.

Acknowledgements

The authors are grateful to the Coordination for the Improvement of Higher Education Personnel (CAPES/Brazil) for financial support.

Author details

Lidiane Sabino da Silva[1], Victor Luis dos Santos Teixeira da Silva[2],
Maria Angélica Simões Dornellas de Barros[1] and Pedro Augusto Arroyo[1*]

*Address all correspondence to: arroyo@deq.uem.br

1 Departament of Chemical Engineering, State University of Maringá, Maringá, Brazil

2 Technology Center, Chemical Engineering Program, Federal University of Rio de Janeiro, Rio de Janeiro, Brazil

References

[1] Intergovernmental Plane on Climate Change. Climate Change 2014 – Mitigation of Climate Change [Internet]. 2014. Available from: http://www.ipcc.ch/pdf/assessment-report/ar5/wg3/ipcc_wg3_ar5_chapter8.pdf [Accessed: 2016-03-11]

[2] International Air Transport Association. Climate Change [Internet]. Available from: https://www.iata.org/policy/environment/pages/climate-change.aspx [Accessed: 2016-03-11]

[3] International Air Transport Association. New IATA Passenger Forecast Reveals Fast Growing Markets of the Future [Internet]. 2014-10-10. Available from: http://www.iata.org/pressroom/pr/Pages/2014-10-16-01.aspx [Accessed: 2016-03-11]

[4] Centro de Gestão de Estudo Estratégicos. Biocombustíveis Aeronáuticos – Progressos e Desafios [Internet]. Available from: https://www.cgee.org.br/documents/10182/734063/biocombustiveis_aeronauticos_24012011_9559.pdf [Accessed: 2016-03-11]

[5] Parker R, Lathoud M. Green aero-engines: technology to mitigate aviation impact on environment. Journal of Mechanical Engineering Science. 2010, 224:529–538. DOI: 10.1243/09544062JMES1515

[6] Velázquez RSG, Kubotani RT, Velázquez SMSG. New fuels for aviation: a case study. Revista Mackenzie de Engenharia e Computação. 2011, 12:77–93.

[7] Moses CA, Roets PNJ. Properties, characteristics, and combustion performance of sasol fully synthetic jet fuel. Journal of Engineering for Gas Turbines and Power. 2009, 131:1–17. DOI: 10.1115/1.3028234

[8] Stocker M. Gas phase catalysis by zeolites. Microporous and Mesoporous Materials. 2005, 82:257–292. DOI: 10.1016/j.micromeso.2005.01.039

[9] Derouane EG. Molecular shape-selective catalysis in zeolites – selected topics. Studies in Surface Science and Catalysis. 1984, 19:1–17. DOI: 10.1016/S0167-2991(09)60078-1

[10] Hu J, Yu F, Lu Y. Application of Fischer-Tropsch synthesis in biomass to liquid conversion. Catalysts. 2012, 2:303–326. DOI:10.3390/catal2020303

[11] Yan Q, Yu F, Liu J, Street J, Gao J, Cai Z, Zhang J. Catalytic conversion wood syngas to synthetic aviation turbine fuels over a multifunctional catalyst. Bioresource Technology. 2013, 127:281–290. DOI: 10.1016/j.biortech.2012.09.069

[12] Dinse A, Aigner M, Ulbrich M, Johnson GR, Bell AT. Effects of Mn promotion on the activity and selectivity of Co/SiO_2 for Fischer-Tropsch Synthesis. Journal of Catalysis. 2012, 288:104–114. DOI: 10.1016/j.jcat.2012.01.008

[13] Khobragade M, Majhi S, Pant KK. Effect of K and CeO_2 promoters on the activity of Co/SiO_2 catalyst for liquid fuel production from syngas. Applied Energy. 2012, 94:385–394. DOI: 10.1016/j.apenergy.2012.02.002

[14] Teiseh EA, Capareda S, Rezenom YH. Cobalt based hybrid Fischer-Tropsch synthesis catalyst for improved selectivity of hydrocarbons in the JP-8 carbon number range from a synthesis gas obtained from the pyrolysis of the MixAlco process derived sludge. Applied Catalysis A: General. 2012, 437–438:63–71. DOI: 10.1016/j.apcata.2012.06.012

[15] Sartipi S, Parashar K, Makkee M, Gascon J, Kapteijn F. Breaking the Fischer-Tropsch synthesis selectivity: direct conversion of syngas to gasoline over hierarchical Co/H-ZSM-5 catalysts. Catalysis Science & Technology. 2013, 3:572–575. DOI: 10.1039/C2CY20744C

[16] Cheng K, Kang J, Huang S, You Z, Zhang Q, Ding J, Hua W, Lou Y, Deng W, Wang Y. Mesoporous beta zeolite-supported ruthenium nanoparticles for selective conversion of synthesis gas to C_5-C_{11} isoparaffins. ACS Catalysis. 2012, 2:441–449. DOI: 10.1021/cs200670j

[17] Ngamcharussrivichai C, Liu X, Li X, Vitidsant T, Fujimoto K. An active and selective production of gasoline-range hydrocarbons over bifunctional Co-based catalysts. Fuel. 2007, 86:50–59. DOI: 10.1016/j.fuel.2006.06.021

[18] Martínez A, Valencia S, Murciano R, Cerqueira HS, Costa AF, Sousa-Aguiar EF. Catalytic behavior of hybrid Co/SiO_2-(medium-pore) zeolite catalysts during the one-stage conversion of syngas to gasoline. Applied Catalysis A: General. 2008, 346:117–125. DOI: 10.1016/j.apcata.2008.05.015

[19] Zola AS, Silva LS, Moretti AL, Fraga AC, Sousa-Aguiar EF, Arroyo PA. Effect of silylation and support porosity of Co/MCM-41 and Co/SiO_2 catalysts in Fischer-Tropsch Synthesis. Topics in Catalysis. 2016, 59:219–229. DOI: 10.1007/s11244-015-0446-1

[20] Gnanamani MK, Pendyala VRR, Jacobs G, Sparks DE, Shafer WD, Davis BH. Fischer-Tropsch synthesis: effect of halides and potassium addiction on activity and selectivity of cobalt. Catalysis Letters. 2014, 144:1127–1133. DOI: 10.1007/s10562-014-1263-z

[21] Girardon J–S, Quinet E, Griboval-Constant A, Chernavskii PA, Gengembre L, Khodakov AY. Cobalt dispersion, reducibility, and surface sites in promoted silica-supported Fischer-Tropsch catalysts. Journal of Catalysis. 2007, 143–157. DOI: 10.1016/j.jcat.2007.03.002

[22] Bezemer GL, Bitter JH, Kuipers HPCE, Oosterbeek H, Holewijn JE, Xu X, Kapteijn F, Dillen AJ, Jong KP. Cobalt particle size effects in the Fischer-Tropsch reaction studied with carbon nanofiber supported catalysts. Journal of the American Chemical Society. 2006, 128:3956–3964. DOI: 10.1021/ja058282w

[23] Yin D, Li W, Yang W, Xiang H, Sun Y, Zhong B, Peng S. Mesoporous HMS molecular sieves supported cobalt catalysts for Fischer-Tropsch synthesis. Microporous and Mesoporous Materials. 2001, 47:15–24. DOI: 10.1016/S1387-1811(01)00322-5

[24] Liu G, Yan B, Chen G. Technical review on jet fuel production. Renewable and Sustainable Energy Reviews. 2013, 25:59–70. DOI: 10.1016/j.rser.2013.03.025

[25] Koroneos C, Dompros A, Roumbas G, Moussiopoulos N. Life cycle assessment of kerosene used in aviation. International Journal of Life Cycle Assessment. 2005, 10:17–24. DOI: http://dx.doi.org/10.1065/lca2004.12.191

[26] Defence Standard 91-91. Turbine fuels, Aviation Kerosene Type, Jet A-1, Nato Code: F-35, Joint Service Designation AFTUR, Ministry of Defence, 2006.

[27] Viljoen CL, Zyl VLV, Tonder EH. Synthetic Aviation Fuel. Pub. No.: US 2010/0264061Al, 2010.

[28] Chevron. Alternative Jet Fuels [Internet]. 2006. Available from: https://www.cgabusinessdesk.com/document/5719_Aviation_Addendum._webpdf.pdf [Accessed: 2016-03-11]

[29] Maurice LQ, Lander H, Edwards T, Harrison WE. Advanced aviation fuels: a look ahead via a historical perspective. Fuel. 2001, 80:747–756, 2001. DOI: 10.1016/S0016-2361(00)00142-3

[30] Dalai AK, Davis BH. Fischer-Tropsch synthesis: a review of water effects on the performances of unsupported and supported Co catalysts. Applied Catalysis, A: General. 2008, 348:1–15. DOI: 10.1016/j.apcata.2008.06.021

[31] Biales JM, Wan YD, Kilpatrick PK, Roberts GW. Separation of Fischer-Tropsch wax from catalyst using near-critical fluid extraction: analysis of process feasibility. Energy Fuels. 1999, 13:667–677. DOI: 10.1021/ef980208u

[32] Dry ME. Present and future applications of the Fischer-Tropsch process. Applied Catalysis A: General. 2004, 276:1–3. DOI: 10.1016/j.apcata.2004.08.014

[33] Westhuizen R, Crous R, Villiers A, Sandra P. Comprehensive two-dimensional gas chromatography for the analysis of Fischer-Tropsch oil products. Journal of Chromatography A. 2010, 1217:8334–8339. DOI: 10.1016/j.chroma.2010.10.099

[34] Miyazawa T, Hanaoka T, Shimura K, Hirata S. Fischer-Tropsch synthesis over a Co/ SiO_2 catalysts modified with Mn and Zr under practical conditions. Catalysis Communications. 2014, 57:36–39. DOI: 10.1016/j.catcom.2014.07.037

[35] Sauciuc A, Abosteif Z, Weber G, Potetz A, Rauch R, Hofbauer H, Schaub G, Dumitrescu L. Influence of operating conditions on the performance of biomass-based Fischer-Tropsch synthesis. Biomass Conversion and Biorefinery. 2012, 2:253–263. DOI: 10.1007/s13399-012-0060-4

[36] Hanaoka T, Liu Y, Matsunaga K, Miyazawa T, Hirata S, Sakanishi K. Bench-scale production of liquid fuel from woody biomass via gasification. Fuel Processing Technology. 2010, 91:859–865. DOI: 10.1016/j.fuproc.2009.09.012

[37] Farias FEM, Sales FG, Fernandes FAN. Effect of operating conditions and potassium content on Fischer–Tropsch liquid products produced by potassium-promoted iron catalysts. Journal of Natural Gas Chemistry. 2008, 17:175–178. DOI: 10.1016/S1003-9953(08)60047-X

[38] Kapteijn F, Deugd RM, Moulijn JA. Fischer–Tropsch synthesis using monolithic catalysts. Catalysis Today. 2005:350–356. DOI: 10.1016/j.cattod.2005.06.063

[39] Dry ME. Practical and theoretical aspects of the catalytic Fischer-Tropsch process. Applied Catalysis A. 1996, 138:319–344. DOI: 10.1016/0926-860X(95)00306-1

[40] Steynberg A, Dry M. Fischer-Tropsch Technology. 1st ed. Netherlands: Elsevier Science; 2004. 722p.

[41] Bao B, El-Halwagi MM, Elbashir NO. Simulation, integration, and economic analysis of gas-to-liquid processes. Fuel Processing Technology. 2010, 91:703–713. DOI: 10.1016/j.fuproc.2010.02.001

[42] Zhang Q, Deng W, Wang Y. Recent advances in understanding the key catalyst factors for Fischer-Tropsch synthesis. Journal of Energy Chemistry. 2013, 22:27–38. DOI: 10.1016/S2095-4956(13)60003-0

[43] Song C, Garcés JM, Sugi Y. Introduction to shape-selective catalysts. In: Song C, Garcés JM, Sugi Y, editors. Shape-Selective Catalysts. 1st ed. Washington: ACS Symposium Series – American Chemical Society; 1999. p. 1–16. DOI: 10.1021/bk-2000-0738.ch001

[44] Bessel S. Investigation of bifunctional zeolite supported cobalt Fisher–Tropsch catalysts. Applied Catalysis A: General. 1995, 126:235–244. DOI: 10.1016/0926-860X(95)00040-2

[45] Zola AS, Bidart AMF, Fraga AC, Hori CE, Sousa-Aguiar EF, Arroyo PA. Cobalt supported on different zeolites for Fischer-Tropsch synthesis. Studies in Surface Science and Catalysis. 2007, 167:129–134. DOI: 10.1016/S0167-2991(07)80120-0

[46] Kibby C, Jothimurugesan K, Das T, Lacheen HS, Rea T, Saxton RJ. Chevron's gas conversion catalysis hybrid catalysts for wax-free Fischer-Tropsch synthesis. Catalysis Today. 2013, 215:131–141. DOI: 10.1016/j.cattod.2013.03.009

[47] Boronat M, Corma A. Factors controlling the acidity of zeolites. Catalysis Letters. 2015, 45:162–172. DOI: 10.1007/s10562-014-1438-7

[48] Corma A. State of the art and future challenges of zeolites as catalysts. Journal of Catalysis. 2003, 216:298–312. DOI: 10.1016/S0021-9517(02)00132-X

[49] Moreno EL, Rajagopal K. Challenges of acidity in solid state catalysis. Química Nova. 2009, 32:538–542. DOI: 10.1590/S0100-40422009000200044

[50] Pine LA, Maher PJ, Wachter WA. Prediction of cracking catalyst behavior by a zeolite unit cell model. Journal of Catalysis. 1984, 85:466–476. DOI: 10.1016/0021-9517(84)90235-5

[51] Barthomeuf D. Zeolite acidity dependence on structure and chemical environment. Correlations with catalysis. Materials Chemistry and Physics. 1987, 17:49–71. DOI: 10.1016/0254-0584(87)90048-4

[52] Bessel S. Support effects in cobalt-based Fischer-Tropsch catalysis. Applied Catalysis A: General. 1993, 96:253–268. DOI: 10.1016/0926-860X(90)80014-6

[53] Ravishankar R, Li MM, Borgna A. Novel utilization of MCM-22 molecular sieves as supports of cobalt catalysts in the Fischer-Tropsch synthesis. Catalysis Today. 2005, 106:149–153. DOI: 10.1016/j.cattod.2005.07.123

[54] Iglesia E. Design, synthesis, and use of cobalt-based Fischer-Tropsch synthesis catalysts. Applied Catalysis A: General. 1997, 161:59–78. DOI: 10.1016/S0926-860X(97)00186-5

[55] Beaumont SK. Recent developments in the application of nanomaterials to understanding molecular level processes in cobalt catalysed Fischer-Tropsch synthesis. Physical Chemistry Chemical Physics. 2014, 16:5034–5043. DOI: 10.1039/c3cp55030c

[56] Davis BH. Fischer-Tropsch synthesis: comparison of performances of iron and cobalt catalysts. Industrial & Engineering Chemistry Research. 2007, 46:8938–8945. DOI: 10.1021/ie0712434

[57] Tijmensen MJA, Faaij APC, Hamelinck CN, Hardeveld MRM. Exploration of the possibilities for production of Fischer-Tropsch liquids and power via biomass gasification. Biomass and Bioenergy. 2002, 23:129–152. DOI: 10.1016/S0961-9534(02)00037-5

Chapter 10

Effects of Natural Zeolites on Bioavailability and Leachability of Heavy Metals in the Composting Process of Biodegradable Wastes

Jiwan Singh, Ajay S. Kalamdhad and
Byeong-Kyu Lee

Additional information is available at the end of the chapter

http://dx.doi.org/10.5772/63679

Abstract

The bioavailability and leachability of heavy metals play an important role in the toxicity of heavy metals in the final compost followed by land application. This chapter examines the effects of natural zeolite on bioavailability of heavy metals (Zn, Cu, Mn, Fe, Ni, Pb, Cd, and Cr) in the form of water soluble and diethylenetriaminepentaacetic acid (DTPA) extractable. The toxicity characteristic leaching procedure (TCLP) test was performed to examine the leachability of heavy metals. Water solubility, DTPA extractability, leachability, and most bioavailable fractions were reduced during agitated pile composting (APC) and rotary drum composting (RDC) of water hyacinth with zeolite addition. The addition of the natural zeolite (clinoptilolite) during the composting process led to an increase in Na, Ca, and K concentrations and significantly reduced the water solubility and DTPA and TCLP extractability of heavy metals. The addition of an appropriate amount of natural zeolite during the composting process enhanced the organic matter degradation, thereby increasing the conversion into the most stabilized organic matter and reducing the bioavailability and leachability of heavy metals.

Keywords: Heavy metals, bioavailability, speciation, natural zeolite, composting, water hyacinth, sewage sludge

1. Introduction

Wastewater treatment plants are increasing worldwide due to urbanization and subsequent increases in sewage sludge production. The management of large quantities of sewage sludge generated from water treatment is a critical problem. Therefore, to overcome the undesirable impacts of the disposal of organic wastes such as water hyacinth and sewage sludge on the environment, it is essential to reduce the volume of these wastes and successfully reuse them as a source of organic matter and nutrients.

Composting is highly an economical method for the handling and final disposal of biodegradable wastes because it is helpful for material recycling and disposal [1, 2]. Water hyacinth (*Eichhornia crassipes*) is a commonly used plant for phytoremediation and constructed wetlands due to its high growth rate and great accumulation of inorganic and organic pollutants [1]. Therefore, the composts prepared from water hyacinth and sewage sludge may contain higher concentrations of heavy metals [3, 4]. The presence of nonbiodegradable and toxic heavy metals in the compost normally prevents its reuse in land applications. If compost with a high concentration of heavy metals is applied to soil, the accumulation of heavy metals in the plants and subsequently the food chain poses a risk to animal and human health [4, 5].

The total concentration of heavy metals measured from compost samples digested with strong acids can indicate the possibility of metal pollution, but cannot specify information related to the bioavailability of heavy metals [5, 6]. The bioavailability of any element specifies the fraction of the element's total content that is easily soluble in an aqueous system, and hence is freely available to plants and soil microorganisms. Water-soluble metals are biologically more dynamic and consequently have a significantly higher risk of contaminating the environment [4, 7]. The diethylenetriaminepentaacetic acid (DTPA)-extractable fraction of metals is a supplemental approach to determine the bioavailable fractions of heavy metals in the compost and soil applied with compost [8]. The toxicity characteristic leaching procedure (TCLP) is applied to evaluate the leaching potential of toxic heavy metals present in waste materials and compost. Heavy metals can be leached from compost and soil amended with compost, and hence pollute groundwater and surface water [9]. The leachability of a metal can be distinguished as the ratio of the quantity of the heavy metal leached from the TCLP test to its total concentration. The TCLP test is commonly used to assess the leachability of heavy metals in compost and whether the compost is hazardous or not [10].

Zeolites are natural hydrated aluminosilicate minerals with a three-dimensional framework structure tetrahedrally coordinated to SiO_4 and AlO_4 [1, 11, 12]. The aluminum ion occupies the center of the tetrahedron of four oxygen atoms. However, the isomorphous replacement of Si^{4+} by Al^{3+} produces a negative charge in the lattice. The Na^+, K^+, and Ca^{2+} are exchangeable for balancing the net negative charge and are commonly exchanged with particular cations (cations of Pb, Th, Cd, Zn, Mn, and NH_4) in aqueous solutions [1, 13]. Natural zeolite has been applied broadly for decreasing the mobility and bioavailability of heavy metals in water hyacinth and sewage sludge composting because of its sorption and exchangeable properties for heavy metals [13, 14]. It can uptake heavy metals that are present in the composting mass in easily available fractions, and exchange with Na and K [15].

2. Composting process

Composting may be defined as a biological breakdown and stabilization of organic substrates, under conditions, which allowed progress of thermophilic temperatures due to heat produced during the degradation of waste biologically. The final product is stable and free from pathogens, and can therefore be applied for land application [16]. **Figure 1** shows the outline of composting process.

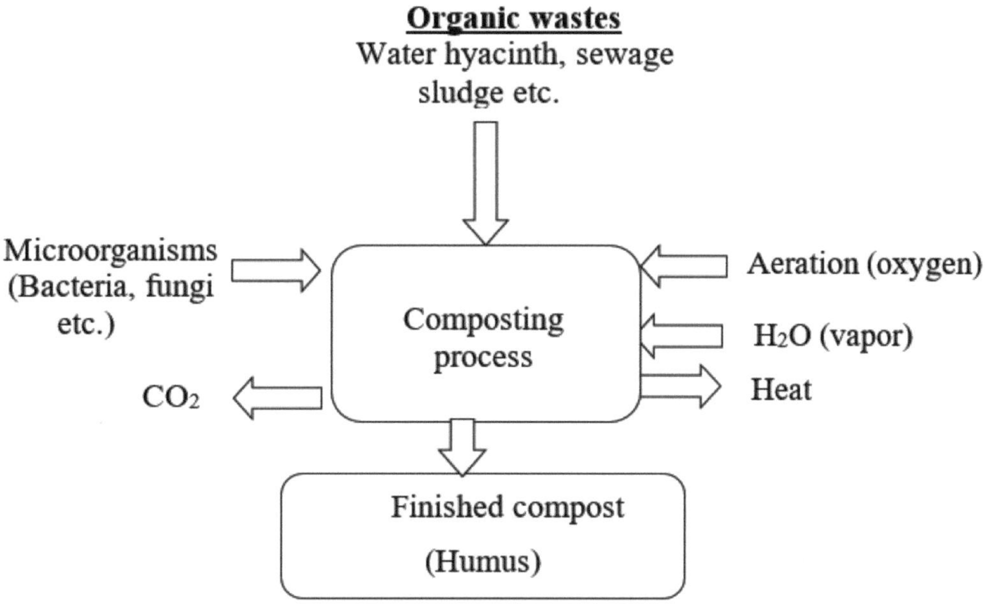

Figure 1. Outline of composting process.

2.1. Factors affecting the composting process

The carbon to nitrogen ratio (C/N ratio) can be a crucial factor in the composting process. To determine the optimal C/N ratio is important for optimizing the composting [17]. C/N ratios between 30 and 50 are favorable for the aerobic composting process. At lower C/N ratio, ammonia can be lost and biological activity is also affected, whereas at higher C/N ratio nitrogen may be a limiting nutrient [18]. The particle size of composting materials should be minimized to ensure efficient aeration and easy decomposition by the bacteria, fungi, and actinomycetes. Therefore, waste materials should be shredded into small pieces between 25 and 75 mm prior to composting [18]. Moisture content is another significant factor affecting the composting process, as moisture greatly disturbs the physical and chemical properties of waste biomass in sequence of the degradation of organic wastes [19]. The moisture content should be in the range between 50 and 60% in the composting process. The optimum moisture content is generally considered approximately 55% [18]. Aeration is an important factor for both microbial growth and gas emission in the composting process [20]. The aeration rate strongly affects microbial activity, substrate degradation rate, and temperature variation in the composting processes of organic wastes [21]). Rasapoor et al. [21] reported that low and medium aeration rates increased the concentration of total nitrogen (TN), which drastically decreased the C/N ratio and lengthened the thermophilic phase. The artificial air supply is generally sustained at 1–2 m^3/day/kg of volatile solids. A temperature increase is an indicator of microbial activities in the process of composting, and thus temperature change can be measured as a suitable parameter to regulate the status of composting processes [22, 23]. Temperature can affect the nature of microorganisms and the rate of decomposition of organic wastes [23]. Lin [24] reported that when the temperature was about 65°C, total coliforms dropped. López-Real and Foster [25] reported that the application of 55°C for only 3–4 days completely eliminated all pathogens. Stentiford [26] stated that the composting temperature must be maintained between 55 and 65°C for the inactivation of total pathogens. The pH is another important parameter for the composting process. The pH can be affected during the composting process due to the production of short-chain organic acids from the feedstock since the early phase of the composting process [27]. The short-chain organic acids and ammonia are maintained in a pH range of 4.9–8.3 [28]. Lower pH decreased microbial activity, thus hindering the growth of composting reaction [27]. However, high pH (>8.5) caused nitrogen loss in the form of ammonia [18].

3. Heavy metals in the composting process

Heavy metals (mg/kg)	Sewage sludge	Municipal solid waste	Water hyacinth
Zn	233	278 ± 22	297.8 ± 3.0
Cu	62	410 ± 26	103.3 ± 0.8
Mn	59.9	–	1105.0 ± 27.5
Fe	3768	–	13300 ± 30
Ni	23	44 ± 7	235.8 ± 1.8
Pb	101	325 ± 24	1537.0 ± 12.5
Cd	0	3.3 ± 0.4	83.8 ± 1.3
Cr	44.03	52 ± 9.2	279.0 ± 1.3

Table 1. Heavy metals concentration (mg/kg) in the final compost of different wastes.

The availability of metals in the final compost is the one of the major sources of soil pollution. Heavy metal pollution in the soil is mainly caused by Cu, Ni, Cd, Zn, Cr, and Pb [29, 30]. Some heavy metals (Fe, Zn, Ca, and Mg) have been reported as having a bioimportance to human being and plants. However, some others (As, Cd, Pb, and methylated forms of Hg) have been reported to have no known bioimportance in human biochemistry and physiology, and their consumption even at very low levels can be toxic to living organisms [31]. **Table 1** illustrates the total concentration of metals (Zn, Cu, Mn, Fe, Ni, Pb, Cd, and Cr) in the final compost of different wastes. The total metal contents are increased in the final compost due to the reduction of organic matter and release of CO_2 during the mineralization process [4]. The total metal concentration found after strong acid digestion of final compost is useful as an overall pollution indicator but provides no useful information about the bioavailable fractions and chemical speciation of metals [5, 6].

4. Heavy metals bioavailability

The bioavailability of heavy metals depends on different extractable fractions rather than on the total metal concentration. Therefore, the bioavailability of heavy metals provides more important evidence of metal toxicity [32]. Even if the heavy metals concentration in sewage sludge or compost is far below the regulation limit, the long-term land application of compost with background heavy metals concentration can increase the content and accumulation of the heavy metals in the soil [10].

Heavy metals in compost and soil amended with compost are commonly separated into two fractions: (i) inert fraction expected as the nontoxic fraction and (ii) the labile fraction, which is supposed to be possibly toxic [33]. To determine the availability of heavy metals, only the labile fraction has been considered bioavailable. The bioavailable fraction can diverge from one metal to another and from one receptor to another. The bioavailability of heavy metals for plants and microorganisms in soil/compost depends on the composition of the different components of soil/compost, such as carbonates, (oxy) metal hydroxides, organic matter, and silica [4, 33].

The heavy metal bioavailability has been considered one of the most critical problematic parameters for the agricultural application of compost [34]. The mobilization of pollutants depends on three factors: their mobility, concentration, and solubility in the compost/soil [33]. The solubility depends on the chemical composition of the leachate in equilibrium with the material; this chemical composition is influenced by the variation of pH that moves the redox equilibrium to predominant forms [33]. Heavy metals are generally present in their hydroxide forms with low solubility at a higher pH. However, at low pH, metals are available in their cationic forms, which are highly soluble and available for plant uptake. There are two types of complex in metal complexation reactions with soil particles: soluble and insoluble. At pH 9, the solubility of Cu is increased due to the formation of soluble complexes [4].

5. Effects of natural zeolite on the bioavailability of heavy metals during the composting process

5.1. Experimental analysis

The details of waste materials are as follows: control (water hyacinth 90 kg + sawdust 15 kg + cattle manure 45 kg), zeolite 5% (control + zeolite 7.5 kg), zeolite 10% (control + zeolite 15 kg), and zeolite 15% (control + zeolite 22.5 kg) during the agitated pile composting (APC) and rotary drum composting (RDC) of water hyacinth [12, 13]. Water-soluble heavy metals were determined after extraction of 2.5 g of sample with 50 ml of distilled water (sample:solution ratio = 1:20) at room temperature for 2 h in a shaker agitated at 100 rpm [4]. DTPA-extractable metals were attained by mechanically shaking 4 g of ground sample (screened through a 0.22-mm sieve) with 40 ml of 0.005 M DTPA, 0.01 M $CaCl_2$, and 0.1 M (triethanolamine) buffered to pH 7.3 at 100 rpm [4]. The standard TCLP method according to the US EPA Method 1311 [35] was applied to the solid samples in order to determine the potential leachability of the heavy metals. A 5 g sample of compost (size less than 9.5 mm) with 100 ml of acetic acid at pH 4.93 ± 0.05 (pH was adjusted by 1 N NaOH) (sample: solution ratio =1:20) was taken in 125 ml reagent bottle and kept at room temperature for 18 h in a shaker at 30 ± 2 rpm. An atomic absorption spectrometer (AAS) (Varian Spectra 55B) was used to analyze Zn, Cu, Mn, Fe, Ni, Pb, Cd, and Cr concentrations in different extracted solutions.

5.2. Total concentration of heavy metals

The total content of heavy metals attained after strong acid digestion is an indicator of compost contamination but cannot provide useful information on the bioavailability of heavy metals in the compost and compost-amended soil [36]. The total concentration of heavy metals was determined during the APC and RDC of water hyacinth with natural zeolite. The total concentration of metals was increased during the APC and RDC process of water hyacinth with natural zeolite (**Table 2**) due to the reduction of organic matter and release of CO_2 during the mineralization processes [12, 13]. In the APC process, the total concentration of Zn, Cu, Mn, Pb, and Cd was increased highest in control as compared to the zeolite treatments; however, the total concentration of Fe, Cr, and Ni was increased highest in 5, 10, and 15% zeolite treatments, respectively [12]. In the RDC process, the total concentration of Cr, Cd, Zn, Mn, Fe, and Pb was increased highest in 5% zeolite treatment. The total concentration of Ni and Cu was increased highest in 10% zeolite treatment and control, respectively [13]. The total concentration of Zn and Ni was increased with increasing amount of natural zeolite addition in the APC process; however, the concentration of these metals was reduced with an increasing amount of natural zeolite in the RDC process of water hyacinth. In the RDC process, the percentage increase of metals was reduced with an increasing amount of natural zeolite due to the reduction in efficiency of the composting process. A higher addition of natural zeolite in the RDC process could hold the high amount of moisture that reduced the microbial activity [13]. Zorpas et al. [37] reported that the heavy metals contents were greatly decreased by approximately 100, 17, 31.7, 35.1, 24.0, 60.0, 56.7, and 47.9% for Cd, Cr, Cu, Fe, Mn, Ni, Pb, and Zn, respectively, in the process of composting of sewage sludge with zeolite. Zorpas et al. [38] concluded that the addition of 25% (w/w) zeolite in compost is adequate to eliminate around 12–60% of the heavy metals during sewage sludge composting. Sprynskyy et al. [14] reported that the addition of 9.09% clinoptilolite to the sludge reduced the total concentration of metals to around 11–51%. The increases in the concentration of heavy metals in the composting process were mainly because of the losses of mass [37].

Composting methods	Days	Total concentration of heavy metals							
		Control	NZ (5%)	NZ (10%)	NZ (15%)	NZ (0%)	NZ (5%)	NZ (10%)	NZ (15%)
		Zn (mg/kg)				Cu (mg/kg)			
APC	0	161.1	277.0	282.5	236.5	31.0	94.8	161.3	134.4
	30	297.8	297.5	306.5	289.7	103.3	128.4	186.4	149.1
RDC	0	147.3	213.9	218.9	240.0	70.5	80.1	88.1	108.8
	20	155.1	290.1	276.0	273.9	91.6	96.7	100.4	115.4
		Mn (mg/kg)				Fe (mg/kg)			
APC	0	573	499	453.3	483	7.92	16.0	21.4	18.5
	30	1105	652.5	780.1	685.5	13.3	30.2	29.0	26.3
RDC	0	538	475	492	531	11.0	19.90	21.1	21.1
	20	668	652	588	620	14.1	31.30	31.5	24.0
		Ni (mg/kg)				Pb (mg/kg)			
APC	0	187.3	225.7	192.0	184.0	872.5	830.0	842.5	832.5
	30	235.8	277.2	243.8	277.5	1537.5	1292.5	1097.5	1004.5
RDC	0	239.5	222.3	244.0	267.3	958	977.0	1037	1032.0
	20	243.5	256.3	285.0	300.7	1111	1209.0	1268	1100.0
		Cd (mg/kg)				Cr (mg/kg)			
APC	0	43.8	56.8	53.3	56.0	257.0	73.0	75.8	87.8
	30	83.8	81.8	68.5	78.5	279.3	115.0	124.3	116.8
RDC	0	51.5	72.4	75.7	75.8	59.3	60.5	76.2	93.9
	20	60.7	91.2	93.9	78.7	76.3	83.5	91.2	105.1

Note: NZ-natural zeolite; control: water hyacinth (90 kg), sawdust (15 kg), cattle manure (45 kg); zeolite 5% (control + zeolite 7.5 kg); zeolite 10% (control + zeolite 15 kg), and zeolite 15% (control + zeolite 22.5 kg).

Table 2. Effects of natural zeolite on concentration of total heavy metals during the water hyacinth composting.

5.3. Water solubility of heavy metals

The water-soluble fraction of heavy metals in the compost was lower than their total concentration, which is the most toxic fraction in the final compost [39]. **Table 3** shows the changes in the water solubility of the heavy metals during the APC and RDC processes. The water solubility of the metals (percentage of total metal) was decreased by approximately 80.0% for Zn, 76.7% for Cu, 83.1% for Mn, and 100% for Cr in the APC process [40]. However, during the RDC process of water hyacinth, the water-soluble fraction of the heavy metals was decreased (percentage of total metal) by approximately 71.3% for Zn, 79.1% for Cu, 78.3% for Mn, 76.8% for Fe, and 89.4% for Cr. The highest reduction of water-soluble Zn, Cu, Mn, and

Composting methods	Days	Water-soluble metals concentration							
		Control	Zeolite 5%	Zeolite 10%	Zeolite 15%	Control	Zeolite 5%	Zeolite 10%	Zeolite 15%
		Zn (mg/kg)				Cu (mg/kg)			
APC	0	3.38	3.59	1.49	1.42	1.12	1.73	1.77	1.60
	30	1.25	0.87	0.46	0.59	0.92	0.55	0.49	0.97
RDC	0	2.28	2.542	1.57	1.368	2.28	1.703	1.825	2.512
	20	1.264	0.991	0.725	0.9125	1.09	0.49	0.435	1.691
		Mn (mg/kg)				Fe (mg/kg)			
APC	0	8.50	10.89	3.80	2.90	16.14	15.28	9.11	8.89
	30	12.68	3.65	1.11	1.29	75.08	9.92	3.65	6.59
RDC	0	9.99	3.29	3.34	1.965	19.32	21.55	31.09	47.79
	20	3.36	0.98	0.965	1.25	20.28	10.715	10.8	37.57
		Cr (mg/kg)				Ni, Pb and Cd (mg/kg)			
APC	0	2.07	0.43	0.62	0.73	ND	ND	ND	ND
	30	0.54	ND	ND	0.38	ND	ND	ND	ND
RDC	0	1.69	1.605	1.725	0.76	ND	ND	ND	ND
	20	0.4	0.235	0.3325	0.64	ND	ND	ND	ND

Note: ND-not detected; control: water hyacinth (90 kg), sawdust (15 kg), cattle manure (45 kg); zeolite 5% (control + zeolite 7.5 kg); zeolite 10% (control + zeolite 15 kg), and zeolite 15% (control + zeolite 22.5 kg).

Table 3. Effects of natural zeolite on water solubility of heavy metals during water hyacinth composting.

Cr was observed in 5% zeolite treatment, whereas a higher reduction of Fe was observed in 10% zeolite treatment [41]. The water-soluble fractions of Ni, Pb, and Cd were not found in the APC and RDC processes. The addition of zeolite significantly reduced the water-soluble Cu, Mn, Fe, and Cr contents during the composting water hyacinth composting process. In the APC process of water hyacinth with natural zeolite, the highest reduction in water-soluble concentration of Zn was found in control. However, the highest reduction in water-soluble fraction of Zn, Cu, Mn, Fe, and Cr was observed in 10% zeolite treatment [40]. In the RDC process, the highest reduction in water-soluble fraction of Zn, Cu, Mn, and Cr was observed in 5% zeolite treatment; however, the highest reduction in Fe was observed in 10% zeolite treatment. The poor reduction in water solubility of metals was observed in 15% zeolite treatment likely due to the higher addition of zeolite, which could hinder the composting process by absorbing water content leading to the clumping of compost biomass [41]. Water-soluble fractions of Ni, Pb and Cd were not found in all zeolite treatments during the APC and RDC processes of water hyacinth [40, 41]. The highest reduction in water solubility of metals was achieved in 10% zeolite treatment during the APC process; whereas in the RDC process of water hyacinth the highest reduction was observed in 5% zeolite treatment. Stylianou et al. [42] reported that during the sewage sludge composting process, decomposed organic matter plays an important role in releasing water-soluble heavy metals, which increases the availability of their exchangeable forms that can uptake by natural zeolite through ion exchange process.

5.4. DTPA extractability

It has been considered that DTPA is a chelating agent and generally used for the analysis of metals availability for plants in the soil amended with heavy-metal-contaminated compost or sewage sludge at regular or even higher concentration [43]. In the case of added zeolite, the amount of DTPA-extractable Pb, Cu, and Zn significantly decreased in the composting because of the higher ion exchange capacity [10]. The DTPA extraction efficiency of the metals was decreased by about 56.6% for Zn, 85.4% for Cu, 81.7% for Mn, 78.5% for Ni, and 75.5% for Cr in the APC of water hyacinth with natural zeolite [40]. However, the DTPA extraction efficiency of metals was decreased by about 58.6% for Zn, 81.1% for Cu, 48% for Mn, 52.1% for Fe, 93.2% for Ni, and 77% for Cr during the RDC process of water hyacinth with natural zeolite [41]. A decrease in DTPA-extractable heavy metals were more found in the RDC process than in the APC process (**Figure 2**). DTPA-extractable concentrations of Pb and Cd were not observed during the APC and RDC processes of water hyacinth with zeolite. In the APC process of water hyacinth, the highest reduction in DTPA extractability of Cu and Mn was observed in control; however, Zn, Cu, Ni, Fe, and Cr were reduced in 5 and 10% zeolite treatments in comparison to the control [40]. The DTPA extraction of Zn, Mn, and Fe was increased in control during the RDC process. However, in the zeolite-added compost, the DTPA extractability of Zn, Cu, Mn, Fe, and Cr was reduced significantly. The reduction in DTPA-extractable heavy metal can be attributed as ion-exchange process where metal cations are mainly exchanged with Na, K, and Ca. DTPA-extractable fraction of Pb and Cd were not found in all zeolite treatments during the APC and RDC processes of water hyacinth [40, 41]. A higher reduction in DTPA extraction of Ni was observed in 10% zeolite treatment followed by 5 and 15% zeolite treatments, and control in the APC process of water hyacinth [40]. A decrease in DTPA-extractable metals in the final compost of water hyacinth with zeolite addition can be attributed to the ligneous bulking agent (sawdust) promoting both the formation and complexation ability of humic acid, resulting reduction in bioavailability of heavy metals [41]. The reduction of DTPA extractability of Zn, Cu, Ni, and Cr was also observed by Chiang et al. [10] during the sewage sludge composting with natural zeolite. In the control test, the concentrations of all tested DTPA metals were higher than those in the zeolite treatments. In matured compost product from the control test, the DTPA solution extracted about 65.0, 18.7, 63.2, and 39.5% of the total Zn, Ni, Cu, and Pb, respectively [10]. DTPA-extractable heavy metal reduction is attributed to the mechanism of ion-exchange processes where metal cations are mainly exchanged with Na, K, and Ca during the composting process [11, 15]. Furthermore, the reduction in DTPA-extractable metals at the end of the composting process due to the transformation of organic matter leads to the formation of metal-humus complexes, which make the metals insoluble and thus less easily extractable [44].

Xiong et al. [45] concluded that ligneous bulking agents, especially wood sawdust, promote both the formation and complexation ability of humic acid, which can reduce the bioavailability of heavy metals, thus reducing the pollution risk of heavy metals in the agricultural application of compost.

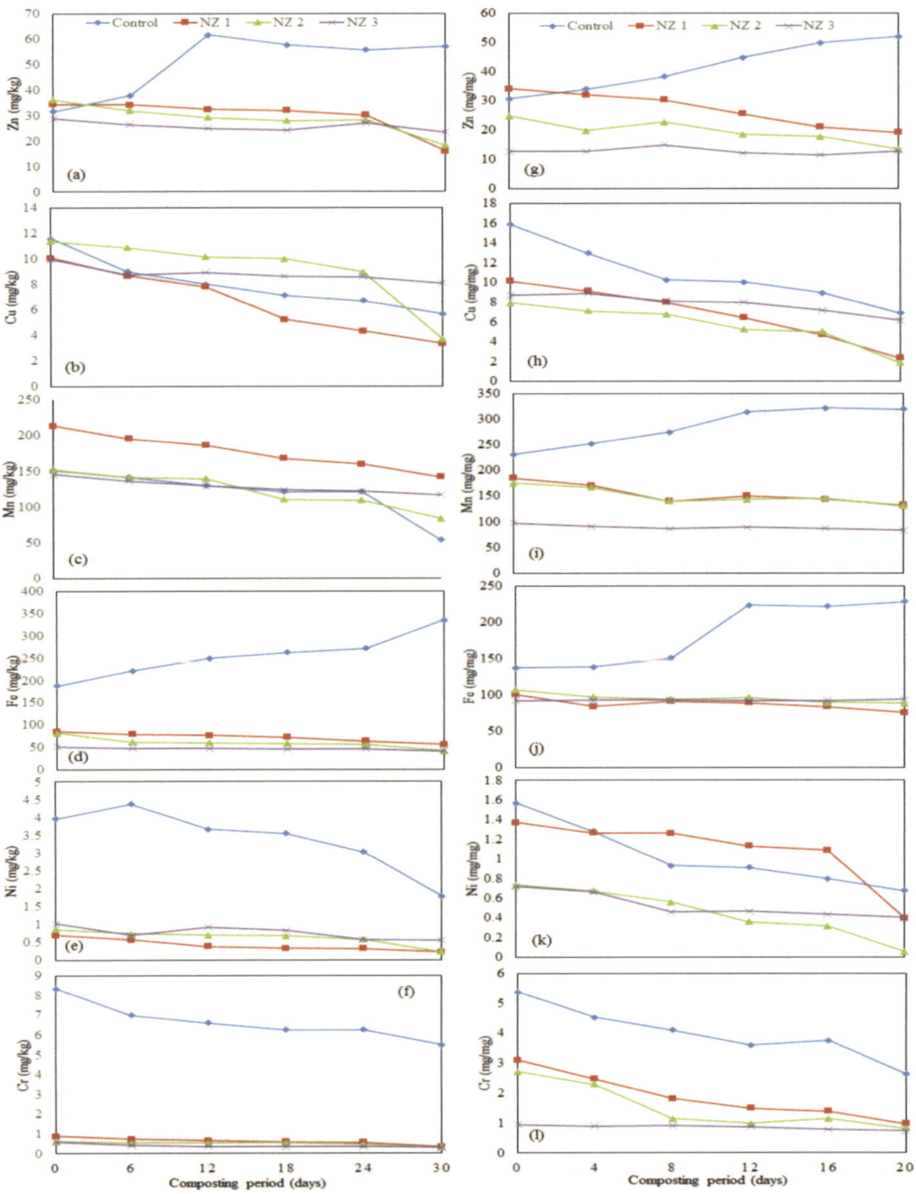

Figure 2. Changes of DTPA-extractable concentration of heavy metals during the agitated pile (a–f) and rotary drum (g–l) composting of water hyacinth (NZ-natural zeolite).

5.5. Leachability of heavy metals

Contaminants	Threshold (mg/kg)
Arsenic	100
Barium	2000
Cadmium	20
Chromium	100
Lead	100
Mercury	4
Selenium	20
Silver	100

Note: Only applicable to samples that are 100% solid.

Table 4. Threshold limits for leachable heavy metals (mg/kg).

The TCLP test is intended to define the mobility of organic and inorganic constituents that are available in liquid and solid wastes [35]. TCLP is used to evaluate the suitability of compost for land application or whether it should be considered a hazardous waste. The procedure is planned to check the leaching possibility of metals in the compost material for agricultural application. The controlling limits for the leached fraction of toxic heavy metals are based on avoiding groundwater pollution through metals, which can create a risk to human health and environment [30]. According to the US EPA [35], the threshold limit for heavy metals contamination in compost is given in **Table 4**. The TCLP-extractable heavy metal concentrations (mg/kg) were in the range of 0.82–3.0, 1.2–3.7, and 12.7–17.9 for Cd, Cr, and Pb, respectively, in the mature compost, confirming that water hyacinth compost was not hazardous to the soil application.

The TCLP-extractable heavy metals were in compliance with the EPA regulatory thresholds limit. **Table 5** shows the variation of leachable heavy metals (Zn, Cu, Mn, Fe, Ni, Pb, Cd, and Cr) during the APC and RDC processes of water hyacinth with natural zeolite. In the APC process of water hyacinth, the TCLP concentration of metals were reduced approximately 61.4% of the total Zn, 72.0% of Cu, 51.4% of Mn, 73.9% of Fe, 64.6% of Ni, 53.3% of Pb, 82.8% of Cd, and 59.7% of Cr in zeolite treatments [40]. In the RDC process of water hyacinth, the leachability of heavy metals was reduced (percentage of total metal) approximately 67.4% for Zn, 52.7% for Mn, 67.0% for Fe, 67.9% for Ni, 71.0% for Cd, and 72.6% for Cr in zeolite treatments. However, the highest reduction in leachability of Cu (73.2%) and Pb (72.4%) was found in the control during the RDC process of water hyacinth [41]. A higher reduction in leachability of metals was observed in zeolite treatments when compared to control in both APC and RDC of water hyacinth [40, 41]. In the RDC process, the highest reduction in leachability of Zn, Mn, Fe, and Ni was observed in 5% zeolite treatment, whereas the reduction of Cd and Cr was observed in 10% zeolite treatment. In the water hyacinth composting with natural zeolite, the pH of the initial feed mixture was enhanced in comparison to control, which reduced the leachability of heavy metals [40, 41, 46]. Furthermore, the reduction in the leachable concentration of heavy metals might be due to humic substances formed at the end of composting process, which had a capacity to form a complex with metals [1, 47]. The reduction in the leachability of Cu and Pb was not significant in any of the zeolite treatments in comparison to control. Increasing the amount of zeolite addition did not reduce the leachability of metals. The reduction in the leachability of heavy metals was much less in 3% zeolite treatment in comparison to control and 1 and 2% zeolite treatments during the RDC process.

Composting methods	**Days**	**Leaching concentration of heavy metals**							
		Control	**Zeolite (5%)**	**Zeolite (10%)**	**Zeolite (15%)**	**Control**	**Zeolite (5%)**	**Zeolite (10%)**	**Zeolite (15%)**
		Zn (mg/kg)				Cu (mg/kg)			
APC	0	30.39	57.53	42.53	37.22	3.47	7.60	12.0	8.0
	30	28.62	23.855	18.86	25.99	3.50	2.88	4.38	3.58
RDC	0	42.22	50.08	38.15	24.77	6.60	13.18	7.31	6.13
	20	21.27	22.20	21.28	20.99	2.30	7.40	3.60	4.45
		Mn (mg/kg)				Fe (mg/kg)			
APC	0	155.10	283.40	259.50	254.0	141.50	53.45	31.02	27.87
	30	233.80	247.90	216.90	210.55	136.70	26.32	16.44	14.99
RDC	0	205.10	184.10	180.40	140.20	121.70	48.30	27.10	32.80
	20	175.50	119.60	130.0	133.90	63.30	27.40	23.40	24.0
		Ni (mg/kg)				Pb (mg/kg)			
APC	0	10.90	17.60	13.42	9.93	36.50	38.50	37.0	36.79
	30	9.50	7.56	7.60	6.84	32.0	280.0	27.0	29.0
RDC	0	8.60	7.69	5.50	5.33	53.20	26.05	22.05	21.40
	20	3.60	2.80	2.10	3.29	17.0	12.87	12.65	17.80
		Cd (mg/kg)				Cr (mg/kg)			
APC	0	1.50	1.98	1.26	1.35	4.62	6.90	5.70	4.30
	30	1.20	0.49	0.43	0.66	3.82	4.38	4.08	2.78
RDC	0	1.86	1.75	2.95	4.55	6.10	2.53	4.30	2.55
	20	1.02	0.82	1.22	3.01	3.65	1.18	1.41	1.65

Note: Control: water hyacinth (90 kg), sawdust (15 kg), cattle manure (45 kg); zeolite 5% (control + zeolite 7.5 kg); zeolite 10% (control + zeolite 15 kg), and zeolite 15% (control + zeolite 22.5 kg).

Table 5. Effects of natural zeolite on leaching concentration of heavy metals during agitated pile and rotary drum composting of water hyacinth.

http://dx.doi.org/10.5772/63679

6. Conclusions

The addition of natural zeolite (clinoptilolite) during the composting process led to significantly reduce the water solubility, and DTPA and TCLP extractability of heavy metals. The TCLP test proved that the concentrations of all selected heavy metals in control and the concentrations of the heavy metals released from the zeolite-treated compost were below the threshold limits. The highest reduction in the bioavailability and leachability of the heavy metals was observed in zeolite treatments 5 and 10% during the APC and RDC processes of water hyacinth. The addition of natural zeolite at suitable concentration successfully reduced the bioavailable and leachable fraction of heavy metals during the composting process of sewage sludge and water hyacinth. Natural zeolite takes up a significant amount of heavy metals during the composting of organic wastes. Addition of the natural zeolite during the composting process led to the increased Na, Ca, and K concentrations and effectively reduced water solubility and DTPA and TCLP extractability of heavy metals. The optimum percentage of zeolite addition in composting mass could fasten degradation of organic biomass; therefore, it decreased the bioavailability and leachability of the heavy metals during the composting process.

Acknowledgements

This work was supported by a National Research Foundation of Korea (NRF) grant funded by the Ministry of Science, ICT, and Future Planning (2013R1A2A2A03013138).

Author details

Jiwan Singh[1], Ajay S. Kalamdhad[2] and Byeong-Kyu Lee[3*]

*Address all correspondence to: bklee@ulsan.ac.kr

1 Department of Environmental Science, Babasaheb Bhimrao Ambedkar University, Lucknow, India

2 Department of Civil Engineering, Indian Institute of Technology Guwahati, Guwahati, India

3 Department of Civil and Environmental Engineering, University of Ulsan, Ulsan, Republic of Korea

References

[1] Villasenor J, Rodriguez L, Fernandez FJ. Composting domestic sewage sludge with natural zeolites in a rotary drum reactor. Bioresour. Technol. 2011; 102(2):1447–1454. doi:10.1016/j.biortech.2010.09.085

[2] Singh J, Kalamdhad AS. Concentration and speciation of heavy metals during water hyacinth composting. Bioresour. Technol. 2012; 124:169–179. http://dx.doi.org/10.1016/j.biortech.2012.08.043

[3] Gupta R, Mutiyar PK, Rawat NK, Saini MS, Garg VK. Development of a water hyacinth based vermireactor using an epigeic earthworm *Eisenia fetida*. Bioresour. Technol. 2007; *98*:2605–2610. doi:10.1016/j.biortech.2006.09.007

[4] Singh J, Kalamdhad AS. Bioavailability and leachability of heavy metals during water hyacinth composting. Chem. Spec. Bioavailab. 2013a; 25(1):1–14. doi: 10.3184/095422913X13584520294651

[5] Walter I, Martinez F, Cala V. Heavy metal speciation and phytotoxic effects of three representative sewage sludge for agricultural uses. Environ. Pollut. 2006; 139:507–514. doi:10.1016/j.envpol.2005.05.020

[6] Venkateswaran P, Vellaichamy S, Palanivelu K. Speciation of heavy metals in electroplating industry sludge and wastewater residue using inductively coupled plasma. Int. J. Environ. Sci. Technol. 2007;4(4):497–504.

[7] Hsu JH, Lo SL. Effects of composting on characterization and leaching of copper, manganese, and zinc from swine manure. Environ. Pollut. 2001; 114:119–127.

[8] Fuentes A, Llorens M, Saez J, Aguilar MI, Marin ABP, Ortuno JF, Meseguer VF. Ecotoxicity, phytotoxicity and extractability of heavy metals from different stabilised sewage sludges. Environ. Pollut. 2006; 143:355–360. doi:10.1016/j.envpol.2005.11.035

[9] Singh J, Kalamdhad AS. Assessment of bioavailability and leachability of heavy metals during rotary drum composting of green waste (Water hyacinth). Ecol. Eng. 2013; 52:59–69. http://dx.doi.org/10.1016/j.ecoleng.2012.12.090

[10] Chiang KY, Huang HJ, Chang CN. Enhancement of heavy metal stabilization by different amendments during sewage sludge composting process. J. Environ. Eng. Manage. 2007; 17(4):249–256.

[11] Yu Y, Shapter JG, Popelka-Filcoff R, Bennett JW, Ellis AV. Copper removal using bio-inspired polydopamine coated natural zeolites, J. Hazard. Mater. 2014; 273: 174–182.

[12] Singh J, Kalamdhad AS. Effects of natural zeolite on speciation of heavy metals during agitated pile composting of water hyacinth. Int. J. Recycl. Org. Waste Agric. 2014; 3(55): 1–17.

[13] Singh J, Kalamdhad AS. Influences of natural zeolite on speciation of heavy metals during rotary drum composting of green waste. Chem. Spec. Bioavailab. 2014;26(2):65–75. doi: 10.3184/095422914X13953177945668

[14] Sprynskyy M, Kosobucki P, Kowalkowski T, Buszewsk B. Influence of clinoptilolite rock on chemical speciation of selected heavy metals in sewage sludge. J. Hazard. Mater. 2007; 149:310–316. doi:10.1016/j.jhazmat.2007.04.001

[15] Zorpas AA, Constantinides T, Vlyssides AG, Haralambous I, Loizidou M. Heavy metal uptake by natural zeolite and metals partitioning in sewage sludge compost. Bioresour. Technol. 2000; 72:113–119.

[16] Haug RT. The Practical Handbook of Composting Engineering. Boca Raton: Lewis publishers; 1993. 1 p.

[17] Adhikari BK, Barrington S, Martinez J, King S. Characterization of food waste and bulking agent for composting. Waste Manage. 2008; 28:795–804. doi:10.1016/j.wasman.2007.08.018

[18] Peavy HS, Rowe DR, Tchobanoglous G. Environmental Engineering. New York: McGraw-Hill; 1985. 661 pp.

[19] Iqbal KM, Shafiq T, Ahmed K. Characterization of bulking agents and its effects on physical properties of compost. Bioresour. Technol. 2010; 101:1913–1919. doi:10.1016/j.biortech.2009.10.030

[20] Guo R, Li GX, Jiang T, Schuchardt F, Chen TB, Zhao, YQ, Shen YJ. Effect of aeration rate, C/N ratio and moisture content on the stability and maturity of compost. Bioresour. Technol. 2012; 112:171–178. http://dx.doi.org/10.1016/j.biortech.2012.02.099

[21] Rasapoor M, Nasrabadi T, Kamali M, Hoveidi H. The effects of aeration rate on generated compost quality, using aerated static pile method. Waste Manage. 2009; 29:570–573.

[22] Tang Z, Yu GH, Liu DY, Xu DB, Shen QR. Different analysis techniques for fluorescence excitation-emission matrix spectroscopy to assess compost maturity. Chemosphere 2011; 82:1202–1208.

[23] Li Z, Lu H, Ren L, He L. Experimental and modelling approaches for food waste composting: a review. Chemosphere 2013; 93:1247–1257.

[24] Lin C. A negative-pressure aeration system for composting food wastes. Bioresour. Technol. 2008; 99:7651–7656.

[25] López-Real J, Foster M. Plant pathogen survival during composting of agricultural waste. In: Gasser JKR, editor. Composting of Agricultural Wastes. London: Elsevier Applied Science; 1985. pp. 291–299.

[26] Stentiford EI. Composting control: principles and practice. In: de Bertoldi M, Sequi P, Lemmes B, Papi T, editors. The Science of Composting, Part 1. Glasgow: Blackie; 1996. pp. 56–59.

[27] Beck-Friis B, Smårs S, Jönsson H, Kirchmann H. Gaseous emissions of carbon dioxide, ammonia and nitrous oxide from organic household waste in a compost reactor under different temperature regimes. J. Agric. Eng. Res. 2001; 78:423–430.

[28] Conghos MM, Aguirre ME, Santamaria RM. Biodegradation of sunflower hulls with different nitrogen sources under mesophilic and thermophilic incubations. Biol. Fertil. Soils 2003; 38:282–287.

[29] Karaca A, Cetin SC, Turgay OC, Kizilkaya R. Effects of heavy metals on soil enzyme activities. In: Sherameti I, Varma A, editors. Soil Heavy Metals, Soil Biology, vol. 19. Heidelberg: Springer; 2010, pp. 237–265.

[30] Chiroma TM, Ebewele RO, Hymore FK. Levels of heavy metals (Cu, Zn, Pb, Fe and Cr) in Bushgreen and Roselle irrigated with treated and untreated urban sewage water. Int. Res. J. Environ. Sci. 2012; 1(4):50–55.

[31] Duruibe JO, Ogwuegbu MOC, Egwurugwu JN. Heavy metal pollution and human biotoxic effects. Int. J. Phys. Sci. 2007; 2(5):112–118.

[32] Nair A, Juwarkar AA, Devotta S. Study of speciation of metals in an industrial sludge and evaluation of metal chelators for their removal. J. Hazard. Mater. 2008; 152:545–553.

[33] Yobouet YA, Adouby K, Trokourey A, Yao B. Cadmium, copper, lead and zinc speciation in contaminated soils. Int. J. Eng. Sci. Technol. 2010; 2(5):802–812.

[34] Gupta AK, Sinha S. Phytoextraction capacity of the plants growing on tannery sludge dumping sites. Bioresour. Technol. 2007; 98:1788–1794.

[35] US Environmental Protection Agency Method 1311—toxicity characteristic leaching procedure (TCLP), 1992. 35 pp.

[36] Cai QY, Mo CH, Wu QT, Zeng QY, Katsoyiannis A. Concentration and speciation of heavy metals in six different sewage sludge-composts. J. Hazard. Mater. 2007; 147:1063–1072.

[37] Zorpas AA, Vassilis I, Loizidou M. Heavy metals fractionation before, during and after composting of sewage sludge with natural zeolite. Waste Manage. 2008; 28:2054–2060.

[38] Zorpas AA, Vassilis I, Loizidou M, Grigoropoulou H. Particle size effects on uptake of heavy metals from sewage sludge compost using natural zeolite clinoptilolite. J. Colloid Interf. Sci. 2002; 250:1–4.

[39] Fang M, Wong JWC. Effects of lime amendment on availability of heavy metals and maturation in sewage sludge composting. Environ. Pollut. 1999; 106:83–89.

[40] Singh J, Prasad R, Kalamdhad AS. Effect of natural zeolite on bioavailability and leachability of heavy metals during rotary drum composting of green waste. Res. J. Chem. Environ. 2013; 17: 26–34.

[41] Singh J, Kalamdhad AS. Uptake of heavy metals by natural zeolite during agitated pile composting of water hyacinth composting. Int. J. Environ. Sci. 2014; 5(1): 1–15.

[42] Stylianou MA, Inglezakis VJ, Moustakas KG, Loizidou MD. Improvement of the quality of sewage sludge compost by adding natural clinoptilolite. Desalination 2008; 224:240–249.

[43] Guan TX, He HB, Zhang XD, Bai Z. Cu fractions, mobility and bioavailability in soil-wheat system after Cu-enriched livestock manure applications. Chemosphere 2011; 82:215–222.

[44] Garcia C, Moreno JL, Hernfindez T, Costa F. Effect of composting on sewage sludges contaminated with heavy metals. Bioresour. Technol. 1995; 53:13–19.

[45] Xiong X, Yan-xia L, Ming Y, Feng-song Z, Wei L. Increase in complexation ability of humic acids with the addition of ligneous bulking agents during sewage sludge composting. Bioresour. Technol. 2010; 101:9650–9653.

[46] Su DC, Wong JWC. Chemical speciation and phytoavailability of Zn, Cu, Ni and Cd in soil amended with fly ash-stabilized sewage sludge. Environ. Int. 2003: 29:895–900.

[47] Wong JWC, Fang M. Effects of lime addition on sewage sludge composting process. Water Res. 2000; 34(15):3691–3698.

Chapter 11

The Influence of Zeolites on Quality Indicators of Soil-Plant Connection and Food Safety

Aleksandra Badora

Additional information is available at the end of the chapter

http://dx.doi.org/10.5772/64590

Abstract

Applying zeolites to natural environment is important from the point of view of monitoring the toxic metals' mobilization (Al(III), Mn(II), Cd(II), and Pb(II)) as well as microelements (Fe, Mn, Zn, and Cu). These elements influenced food chain and their deficiency as well as excess and determined plant quality and health of humans and animals. Furthermore, zeolites, while having particular physical, physicochemical, and chemical properties, interact with physical, physicochemical, chemical, and biological features of the soil and may lead to alterations in their properties. This exchange is dependent on many factors, i.e. pH, concentration of metal ions in solution. When natural zeolites are applied, one should bear in mind that they are ecological material and do not show any harmful action neither to humans nor to animals. Studies in this chapter will show the influence of described and tested zeolites on the properties and quality indicators of the first food chain link: soil-plant as well as on the quality of food. It would allow to understand, predict, and control the behavior of these elements in natural environment as well as evaluate their potential toxicity and bioavailability.

Keywords: heavy metals, zeolites, some soil properties, some element ratios in plants, food security

1. Introduction

The primary effect of the negative impact of acidic reaction consists in adverse changes in physical, chemical, and biological properties of soils, as well as poor growth and development of plants (lower yields). Secondary effect is the mobilization of aluminum and heavy metal ions in amounts proportional to the acidity of the soil. In acidic soils with pH below 4.2, aluminosilicates are decomposed, whereas the concentration of Al^{3+} and Mn^{2+} ions that occupy space of Mg^{2+} and Ca^{2+} in the sorption complex and contribute to increased leaching of alkaline cations is increased. These changes are a major cause of poor growth and development of crops [1–4]. At first, high concentration of aluminum ions has a destructive effect on the root system, which makes nutrient uptake difficult and results in Mg and P deficiency symptoms in plants and their poorer growth. Aluminum ions also limit the growth of shoots leading to the decay of vertices and the necrosis of leaves. Aluminum also forms the ion ratios in cereals, including enhancement of K:(Ca + Mg) ratio and affects the higher content of Mn and Fe in the above-ground parts of plants. Finally, toxic aluminum reduces the quantity and quality of cereal crops [2, 3, 5]. Considering the influence of environmental conditions on the quality of wheat grain, the health aspect of raw material cannot be overlooked, especially the content of heavy metals, including cadmium, even small amounts of which are harmful to the health of humans and animals and which is characterized by very high mobility within the environment. Cereals are the main source of cadmium for humans.

There is quite broad differentiation in the quantity of cadmium uptaken from the soil. More cadmium is absorbed by durum rather than common wheat kernels [6, 7]. Diverse cadmium levels are also found in grains depending on the cultivar [8–11]. The presence of heavy metals, such as Pb and Cd, in the soil, significantly affects the quality of grain yield, since exceeding the permissible content of these elements in the intervening purchase of cereals (>0.2 mg/kg) disqualifies the grain for further processing and consumption, which may lead to substantial economic losses. It is therefore necessary to monitor the heavy metals content in soils [12, 13].

Applying new sorbents, such as zeolites, in natural environment is important from the point of view of monitoring the toxic metals mobilization (Al, Mn, Cd, and Pb) as well as microelements (Fe, Mn, Zn, and Cu) or macroelements (Mg, Ca, and K). These elements influenced food chain and their deficiency as well as excess and determined plant quality and health of human and animals [4, 5, 14]. Furthermore, new substances, while having particular physical, physicochemical, and chemical properties, interact with physical, physicochemical, chemical, and biological features of the soil, its phases (solid, liquid, and gaseous), as well as its components (mineral-organic complex, soil solution) and may lead to alterations in their properties [14–18]. This exchange is dependent on many factors, i.e. pH, concentration of metal ions in solution, and their speciation. The transport of inns in the environment begins in the liquid phase. This pathway is not only the manner in which ions translocate but also has significant importance for metal accumulation in plants and the food chain [10, 11, 19–21]. Therefore, the knowledge of the properties and behavior of new substances in particular elements of a natural environment does not only have cognitive, but also social aspects.

Among zeolites, clinoptilolite is the most common and the best studied one, not only due to its largest abundance in nature and lowest prices, but also the widest scope of its physicochemical properties [22]. This zeolite is characterized by the following features: large sorption and ion-exchange capacity, ion-exchange selectivity, properties of molecular sieve, catalytic activity, structural thermal stability at temperatures of up to 700–750°C. The chemical composition of zeolites determines their behavior under particular conditions. In an alkaline environment, these minerals are decomposed—their fibrous crystals collapse—mordenite converts into analcites. In an acidic environment, a zeolite can be completely deprived of cations, which in consequence produces a hydrated amorphous silica skeleton that retains its initial shape of crystal. Natural clinoptilolite is characterized by Si/AI ratio >4 as well as higher contents of Na^+ and K^+ cations in relation to Ca^{2+}, Ba^{2+}, and Sr^{2+}. When natural zeolites are applied, one should bear in mind that they are ecological material and do not show any harmful action neither to humans nor to animals. Therefore, zeolites found quite wide spectrum of application as additives to feed and fertilizers [23], as well as soil conditioning agents [24–26]. Ion-exchange ability and sorption capacity of natural zeolites can be used at gradual and uniform release of nutrients into the soil, which prevents against their quick elution [27]. Applying zeolites to the production of mineral fertilizers allows a gradual and controlled introduction of such necessary nutrients as potassium ammonium or phosphates into the soil. Moreover, fertilizers containing some addition of zeolites modified with Cu(II) ions can be applied to copper-deficient soils. Thus, it is possible to reduce the soil irrigation, because water is well retained within zeolite's structure, as well as to decrease soil acidity [2–4, 8, 28] and to reduce the temperature oscillations. These sorbents can also be effectively used as pesticide and herbicide carriers [29].

The group of goals, which the author of this chapter proposed to describe are based on own research and literature and focus on the influence of zeolites on the following quality indicators: changes in the physicochemical properties of certain soils (pH, sorption capacity, bases saturation degree, and amounts of Cd and Pb in soils), certain food quality factors connected to toxic substances and elements (ratios of Fe:Mn and K:Ca + Mg) as factors in plants (wheat and rape).

2. Methodology

The experimental unit during incubation studies consisted of a pot containing 0.5 kg of air-dried soil weight of natural origin, from the acidity soil class considered as acidic soils (pH range 4.5–5.5). The soil moisture content will be maintained at the level of 60% of the field water capacity, optimum for a proper growth and development of test plants. Common wheat (*Triticum aestivum* L.) of Opatka cv. was the test plant used for the soil incubation experiments. Twelve seeds will be sown in each pot. After emergence (KD = 11, according to Zadoks), the number of plants will be reduced to six. Plants will be harvested at full ripeness. Rape biomass was harvested at the shooting stage (KD = 20), according to Zadoks—grade proposed by the European Association for Plant Breeding. Variants of Pb experiment were composed by introducing lead salts (in the form of $Pb(NO_3)_2$) in the amount of 100 mg Pb/kg soil, immobilizing agents such as zeolites—clinoptilolite (Fluka) at the following rates: zeolite-1—300 mg/kg soil, zeolite-2—600 mg/kg soil. Variants of Cd-experiment were composed by introducing cadmium salts (in the form of $CdCl_2$) in the amount of 5 mg Cd/kg soil, immobilizing agents such as zeolites—clinoptilolite (Fluka) at the following rates: zeolite-3—15 mg/kg soil, zeolite-4—30 mg/kg soil. Both plants were growing in both Pb and Cd experiments.

The soil material was subjected to the following analyses: pH in 1 mol KCl /dm^3 at soil-solution ratio equal to 1:2.5, hydrolytic acidity—Hh, contents of exchangeable metal forms (Cd, Pb) in soils was determined in 1 M HCl at soil-solution ratio equal to 1:10. The extraction was supported by intensive mixing using rotational mixer for 1 h. Measurements will be made by means of atomic absorption spectrometry using the Hitachi Z-8200 device with Zeeman's polarization. The content of exchangeable ions (K, Na, Ca, and Mg) was also determined by means of extraction in 1 M CH_3COONH_4 and AAS determination (as above).

The plants' material was collected after harvesting. After drying at 70°C, plant samples were digested in concentrated H_2SO_4 with the addition of 30% H_2O_2 to accelerate the process. In achieved mineralized solutions, determinations of Fe and Mn ions were made. In addition, the following items were calculated: the sum of exchangeable alkali ($S = Na^+ + K^+ + Ca^{2+} + Mg^{2+}$), sorption capacity ($T = Hh + S$) expressed in mmol (+)/kg of soil as well as the alkali saturation degree ($V = S/T \times 100\%$), molar ratios of Fe:Mn and K:(Ca + Mg) for investigated plants as food quality factors, the achieved numerical data were processed statistically applying variance analysis with Tukey confidence semi-intervals at the significance level of 0.05.

3. Zeolites and some soil condition

Zeolites exerted their strongest influence on the pH values increase in the soil contaminated with lead compounds, namely at higher doses (**Figures 1** and **2**). During the second experimental year (under rapeseed), soil pH values decreased and the soil reaction became acidic instead of slightly acidic with oscillations around 4.5–5.5 (**Figures 1** and **2**). Many authors observed different impact of zeolites on the change in pH values [1–3, 7, 29] and on the decrease in mobility of various heavy metals in soils. The reactions could be also dependent on cadmium and lead chemistry, which is connected with the strength of alkalization of both elements [7, 10, 17, 18].

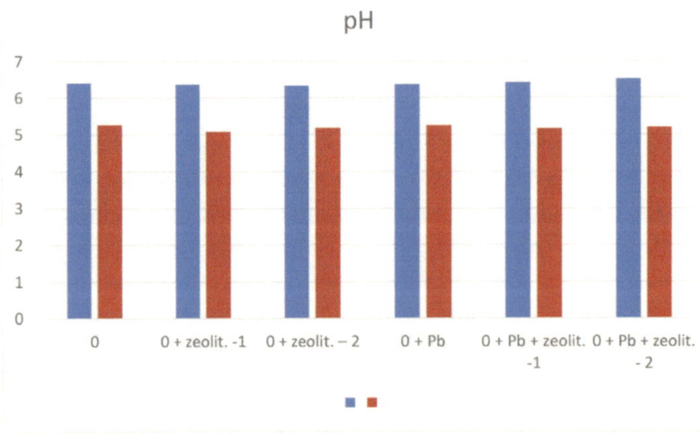

Figure 1. The pH values of soils with Pb and zeolites—own results—(blue color – under wheat, red color – under rapeseed).

Higher levels of the total sorption capacity were recorded in variants with the addition of zeolites in the presence of both heavy metals (**Tables 1** and **2**). It was found that, in comparison with the first year of experiments, the *T* value increased in the second year by over 8 mmol (+)/kg. The base saturation degree (*V*) was within the range of optimum limits and for Polish soils did not exceed 90%. The parameters of *V* were significantly differentiated between the tested plant species in both experimental years. The value of *V* indicators was found to be higher by about 10% in the first year of experiments as compared with the second year (**Tables 1** and **2**). A higher increase was also recorded for the variant with a double dose of zeolites (0 + Cd + zeolites-4) (**Table 2**).

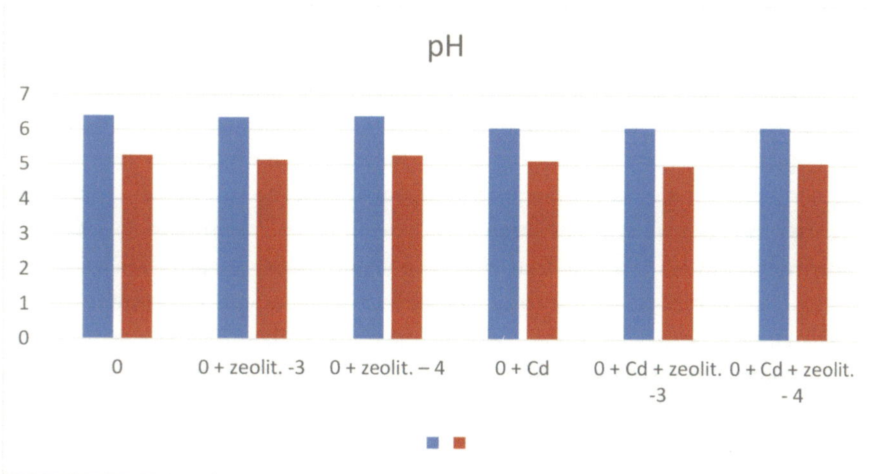

Figure 2. The pH values of soils with Cd and zeolites—own results—(blue color – under wheat, red color – under rapeseed).

No.	Variants (A)	T mmol (+)/kg Under wheat (B)	Under rapeseed (B)	V (%) Under wheat (B)	Under rapeseed (B)
1	0	57.03	71.12	80.72	70.99
2	0 + zeolites-1	59.01	73.07	81.83	68.94
3	0 + zeolites-2	57.40	76.11	81.68	70.93
4	0 + Pb	63.12	77.60	85.15	72.58
5	0 + Pb + zeolites-1	67.90	79.40	85.89	71.90
6	0 + Pb + zeolites-2	69.53	79.45	87.22	71.69
$LSD_{0.05}$	A	8.45		7.55	
	B	4.48		1.55	

Table 1. Total sorption capacity (*T*) and bases saturation degree (*V*) of investigated soils with Pb (own results).

No.	Variants	T mmol (+)/kg		V (%)	
	(A)	Under wheat	Under rapeseed	Under wheat	Under rapeseed
		(B)	(B)	(B)	(B)
1	0	57.03	71.12	80.72	70.99
2	0 + zeolites-3	57.98	75.21	80.79	69.35
3	0 + zeolites-4	59.29	76.61	80.81	72.33
4	0 + Cd	70.50	78.37	81.12	69.55
5	0 + Cd + zeolites-3	69.03	79.98	82.64	69.00
6	0 + Cd + zeolites-4	73.19	82.55	83.57	71.13
$LSD_{0.05}$	A	9.01		6.72	
	B	4.91		1.29	

Table 2. Total sorption capacity (*T*) and bases saturation degree (*V*) of investigated soils with Cd (own results).

Here, the presented results of the experiments revealed that the soil pH is largely determined by the presence of zeolites and also changes in toxic element contents in soil. Characteristics of the soil sorption complex indicate that applying immobilizing agents had an impact on the improvement of its properties through an increase in total sorption capacity (*T*) and bases saturation degree (*V*), both in variants with no heavy metals and lead-contaminated ones, which may be contributed to soil pH increase and, in consequence, a lower share of acidic cations. It was mainly zeolites at their double doses that affected the increase in total soil sorption capacity, which was also probably the consequence of pH changes. Those results could even be confirmed by other authors [22, 29]. Numerous authors [2, 7, 14, 16, 17] have reported that minerals in the soil contain permanent charges, while sorption capacity increases along with soil pH value, mainly due to the dissociation of H^+ and Al^{3+} ions originating from these permanent charges on mineral fragments of the sorption complex. To improve the physicochemical properties of soil is one of the basic conditions that sorbents used for the detoxication of soils contaminated with heavy metals should meet [21].

No.	Variants	Pb		Variants	Cd	
		Wheat grain	Rapeseed biomass		Wheat grain	Rapeseed biomass
1	0 + Pb	0.80	1.78	0 + Cd	2.98	32.85
2	0 + Pb + zeolites-1	0.85	3.40	0 + Cd + zeolites-3	2.82	23.70
3	0 + Pb + zeolites-2	0.70	3.00	0 + Cd + zeolites-4	2.74	27.83
	LSD 0.05	0.16	3.71	LSD 0.05	2.80	14.23

Table 3. The content of Pb and Cd in the investigated plants grown on soils with Pb and zeolites (own results).

The results revealed that applying sorbents contributed to a decrease in mobile lead ion concentrations in soil. In the soil under wheat, lead ion detoxication was observed after introducing a lower zeolite dose, whereas higher rates of the zeolite appeared to be the most efficient Pb^{2+} immobilizing agent in the soil under rapeseed. On the other hand, numerous authors [10, 17, 21] have reported the lower affinity of cadmium, as compared to lead ions, toward zeolites; here, presented results did not reveal that cadmium can be equally sufficiently bound by those minerals (**Table 3**).

It is necessary to emphasize that the reduction in cadmium concentration in both plants was observed (**Table 3**), although the pH in soil was almost at the same level in all variants with this element and zeolites—at around 6 under wheat and around 5 under rape (**Figures 1** and **2**). Such a clear effect was not observed in the Pb reduction. Many authors [4, 9, 28] have shown that zeolites can decrease cadmium toxicity more than other heavy metals. Other authors [5, 30] noted that rape and wheat showed different reactions to different heavy metal toxicity levels.

The increase in the cation sorption capacity is strongly associated with the decrease in toxic metals mobility within the soil environment. Increase in the total sorption capacity index results from the introduction of a material containing functional groups. These materials are clay minerals. Clay minerals having a larger number of pH-dependent sites (e.g. montmorillonite) are more important for heavy metals sorption, as opposed to those producing more pH-independent sites, e.g. kaolinite. Metal—solid phase in the soil bindings formed due to nonspecific sorption manifests a weaker character of bonds with the presence of water molecule between the metal and the adsorbent. Numerous studies revealed the immobilizing effect of zeolites, both natural and artificial, on toxic elements. Synthetic zeolites of the 13X and 4A types were found to reduce the availability of Pb by 70%, Cu by 57%, Ni by 53.5%, Zn by 67.5%, and Cd by 61%, as well as Pb content in tissues of certain plant species [21]. It was also reported that the addition of a natural zeolite, clinoptilolite, had positive effects on the increase in cation-exchangeable capacity (CEC) of the soil [2, 3, 21–23].

4. Zeolites and certain elements' ratios in plants as food security factors

The concept of culture (fertility) of soil means such a condition of the soil which provides growing plants with a sufficient amount of nutrients, water, and air, and is the result of many natural and soil-forming factors dependent on the climate, bedrock, and vegetation. The highest cereal yields can be usually obtained from soils abundant in high cultures (with an optimum content of available forms of plant nutrients such as Mg, K, Ca, P, N, and humus), with regulated relations between water and air and with the pH value close to neutral (pH 5.6–7.2) [13].

The present study revealed the influence of zeolites on the changing of investigated ratios such as Fe:Mn and K:Ca + Mg) (**Figures 3** and **4**). The presence of zeolites influenced positive regulation of Fe:Mn ratios in both investigated plants in the variations with Cd in soil (**Figure 4**). In the variations with Pb in soil (**Figure 3**) optimal values of Fe:Mn ratios were obtained only for wheat grain, not for rape, where lower manganese uptake was observed.

Many studies [6, 8, 9, 14] have shown that the varied concentration of heavy metals has substantial influence on other elements' uptake by plants.

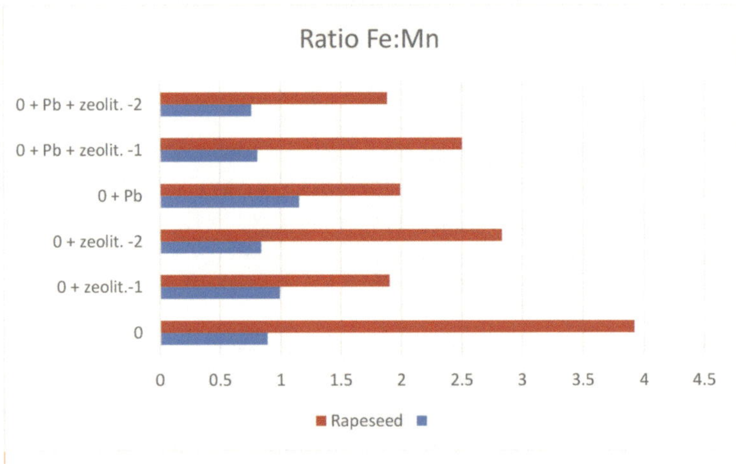

Figure 3. The influence of zeolites used for soils polluted with Pb on Fe:Mn ratio changing in wheat grain (blue color) and rapeseed biomass (red color)—own results.

The content of microelements in soils and their availability to plants depends on many factors [2–4]. Manganese deficit impairs metabolic functions of plants and reduces the sowing value of seeds. Feeding plants of winter and spring wheat with microelements has positive effects on the features of grain quality such as gluten content and sedimentation rate [4, 13, 31, 32].

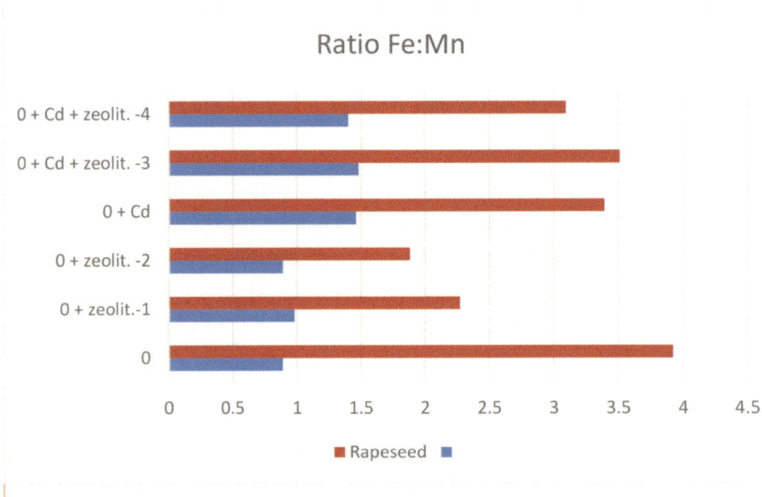

Figure 4. The influence of zeolites used for soils polluted with Cd on Fe:Mn ratio changing in wheat grain (blue color) and rapeseed biomass (red color)—own results.

In the case of K:(Ca + Mg) ratios, the presence of zeolites for the Pb and Cd immobilization, a stronger positive change was observed for wheat grain than for rape (**Figures 5** and **6**). Many authors [4, 9, 14] have shown that the optimal presence of K and Ca and Mg elements in raw materials is very important for food quality. The presence of undesirable substances in plant material entails a risk to consumer's health. The security of crop production is therefore closely linked to the status of the natural environment. Certain substances used a few dozen years ago have been interacting with the environment and the trophic chain until now, being subject to its continuous bioaccumulation [13, 20, 31]. The Polish production fields are characterized by the lack of so-called "elements of life" (Mg, Se). Magnesium deficiencies in the area of agricultural production result in deficiencies of these elements in plant material. Using zeolites for Cd and Pb immobilization in soils may also change the presence of K, Mg, and Ca in plants (**Figures 5** and **6**). The elements discussed above become parts of the food chain mainly through soils: solid surface (soil) attracts and retains the ions, atoms or the molecule layer. The sorption abilities of the soil result from the properties of the sorption complex made of soil colloids: clay minerals (smectites, vermiculite, illite, and kaolinite), crystalline and amorphous iron and aluminum oxides, amorphous minerals, humus, and clay-humus complexes [15–17, 19, 26].

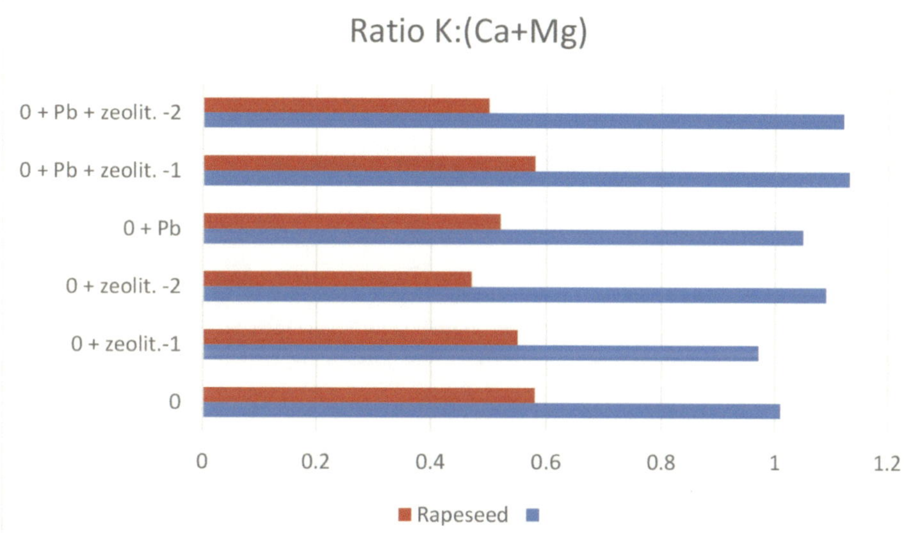

Figure 5. The influence of zeolites used for soils polluted with Pb on K:(Ca + Mg) ratio changing in wheat grain (blue color) and rapeseed biomass (red color)—own results.

Food of either plant or animal origin (sometimes also mineral) is consumed in order to provide the body with energy and nutrients. The basic unit of energy understood in this sense is kilocalorie. Such a measurement system makes it possible to estimate the amount of energy needed by the human body to regenerate itself. An insufficient amount of energy and a low intake of calories lead to hunger and subsequently to death. The daily number of calories required by a person depends on age, sex, body weight, type of work performed, and climate. The World Health Organization (WHO) has established that an adult person requires a minimum of 2200 cal/day to survive. The effects of qualitative malnutrition are not easily noticeable. People affected by it may have a normal body weight and still suffer from the effects of qualitative malnutrition. Vitamin and mineral salts deficiency can lead to serious health issues, such as significantly greater susceptibility to infectious diseases, loss of vision, anemia, coma, reduced knowledge acquisition skills, intellectual development disorder, various forms of physical deformities, and finally death. The most common deficiencies involve the following three elements: vitamin A, iron, and iodine [4, 12, 13, 20, 29, 31, 32].

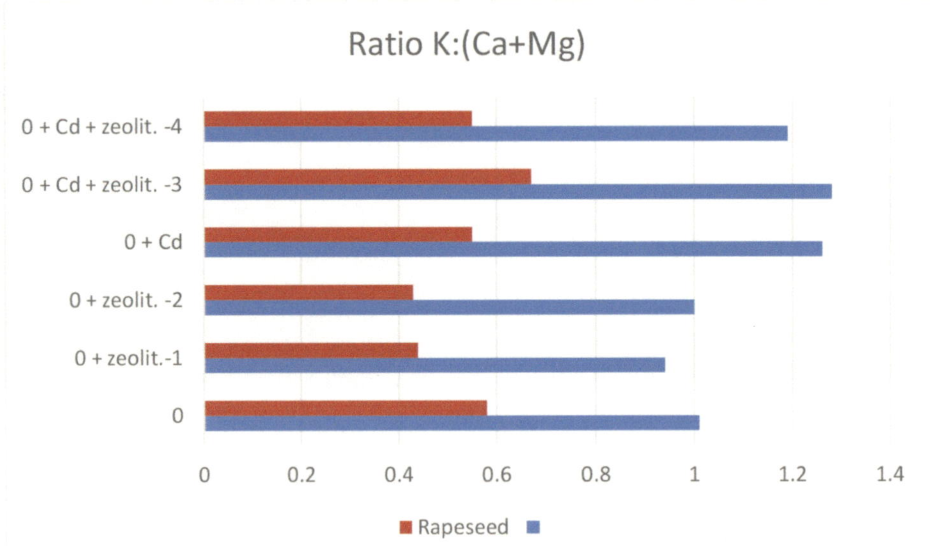

Figure 6. The influence of zeolites used for soils polluted with Cd on K:(Ca + Mg) ratio changing in wheat grain (blue color) and rapeseed biomass (red color)—own results.

5. Summary

It is important to know that half the people suffering from micronutrient deficiency are usually afflicted by cumulative deficiencies of elements, i.e. they are simultaneously lacking several vitamins and minerals in their diet. Qualitative malnutrition is a direct or indirect cause of half of deaths among children under five in the world. As a consequence of deliberate intervention in agro-ecosystems, a man can control their productivity and increase the amount of produced biomass, which can be utilized as food for humans, feed for animals, and raw material for many industry branches. Future of the agriculture should therefore be based on a variety of plants species, from which new, healthier, and less processed goods are made. Such approach promotes not only the safety, but also the nutritional sovereignty of societies.

Zeolites have appeared to be promising binding agents for lead ions mobility and for cadmium at lower doses. Also soil pH values have been changing due to applied sorbents, largely determining the forms of toxic metals in soils. Zeolites improved properties of the soil sorption complex through the increase of total sorption capacity as well as base saturation degree, which met the necessary condition of the lack of toxicity to immobilization of the heavy metals contaminated soils. Zeolites influence also the changing of Fe:Mn and K:(Ca + Mg) ratios in the investigated plants like wheat and rape. These factors can describe the quality of food products made from these plants.

Author details

Aleksandra Badora

Address all correspondence to: aleksandra.badora@up.lublin.pl

Department of Agricultural and Environmental Chemistry, Subdepartment of Quality and Standardization of Plant Materials, University of Life Sciences in Lublin, Lublin, Poland

References

[1] Badora A., Furrer G., Grunwald A., Schulin R., 1998. Immobilization of zinc and cadmium in polluted soils by polynuclear Al_{13} and Al-montmorillonite. J. Soil. Cont. 7, 573–588. DOI: 10.1080/10588339891334447.

[2] Badora A., 2011. Mineral sorbent in natural environment: chosen topics. University of Life Sciences Press, pp. 63. ISBN 978-83-7259-199-9.

[3] Badora A., Hubicki Z., Filipek T., 2011. Action of two different zeolites on chosen experimental bottom properties and some chemical indicators of plants. Przem. Chem., 9, 1779–1783. ID: YADDA bwmeta1.element.baztech-8502ba2a-75c3-4553-9eda-56d0094259d5.

[4] Badora A. (Ed.), 2012. Quality cultivation and standardization of vegetable stocks. University of Life Sciences Press, Lublin, ss. 270.

[5] Marschner H., 1998. Mineral nutrition of higher plants. 2nd edition. Academic Press, Harcourt & Company, Publishers, London, San Diego, New York, Boston, Sydney, Tokyo, Toronto, pp. 680.

[6] Norvell A.W., 1988. Inorganic reaction of manganese in soils. In: Graham R.D., Hannam R.J., Uren E.C. (Eds.), Manganese in soils and plants. Kluwer Academic Publishers, Dordrecht, pp. 37–58.

[7] Badora A., 2012. Influence of zeolites, humic acids, and selenates (VI) on lead and cadmium immobilization and selected soil properties. Polish J. Environ. Stud., 21(4), 813–820. ISBN: 1230-1485.

[8] Badora A. 2001. Aluminum and manganese mobility in the soil. Polish J. Soil Sci., 34, 1–8. ISSN: 0079-2985.

[9] Badora A., 2002. Bioaccumulation of Al, Mn, Zn and Cd in pea plants (*Pisum sativum* L.) against the background of unconventional binding agents. Polish J. Environ. Stud., 11, 109–116. ISBN: 1230-1485.

[10] Alloway J.B., 1990. Heavy metals in soils. Halsted Press, Blackie, Glasgow and London.

[11] Alloway B.J., Ayres D.C., 1999. Chemical grounds of environmental pollution. PWN, Warszawa, pp. 423.

[12] Ziegler J., 2013. Destruction massive. Geopolitique de la faim. Copyright by Instytut Wydawniczy Książka i Prasa, Warszawa, 2013, pp. 369. (in Polish).

[13] Badora A.,Kozłowska-Strawska J., Domańska J., Filipek T., 2014. Cereals – health or disease. Probl. Sustain. Dev., 9(2), 87–98. ID: YAADA bwmeta1.element.baztech-9436df4d-ce74-4da2-8669-6cc1d21fde8a.

[14] Robson A.D., 1989. Soil acidity and plant growth. Academic Press, Australia. Harcourt Brace Jovanovich, Publishers, pp. 1–306.

[15] Sparks D.L., 1995. Environmental soil chemistry. Academic Press, San Diego, New York, Boston, London, Sydney, Tokyo, Toronto.

[16] Sposito G., 1989. The chemistry of soils. Oxford University Press, Oxford, England.

[17] McBride B.M., Martinem E.C., Topp E., Evans L., 2000. Trace metal solubility and speciation in a calcareous soil 18 years after no-till sludge application. Soil Sci. 165, 646–656.

[18] Appelo C.A.J., Postma D., 1994. Geochemistry, groundwater and pollution, AA. Balkema Rotterdam, Brookfield, pp. 536.

[19] McBride B.M., 1994. Environmental chemistry of soil. Oxford University Press, New York, Oxford, 490 ss. 406.

[20] Coultate T.P., 2002. Food. The chemistry of the components. RSC Publishing, UK, pp. 432.

[21] Usman, A.R.A., Kuzyakov Y., Lorenz K., Spahr K., 2006. Remediation of a soil contaminated with heavy metals by immobilizing compounds. J. Plant Nutr. Soil Sci., 169, 205–212. DOI: 10.1002/jpln.200421685.

[22] Pitcher S.K., Slade R.C.T., Ward N.I., 2004. Heavy metal removal from motorway stormwater using zeolites. Sci. Total Environ., 161–166. PMID: 15504502.

[23] Leung S., Barrington S., Zhao X., El-Husseini B., 2006. Effect of particle size on physiochemical properties of clinoptylolite as feed additive. Microporous Mesoporous Mater., 95, 48–56. DOI: 10.1016/j.micromeso.2006.04.010.

[24] Toma M., Hiradate S., Saigusa M., 1999. Chemical species of Al in a gypsum-treated Kitakami Andosol. Soil Sci. Plant Nutr., 45, 279–285. DOI: 10.1080/00380768.1999.10409343.

[25] Tomoshiro M., Yukihiko N., 1999. A new antibacterial agent: antibacterial zeolite. Artif. Organs, 23, 129–130. DOI: 10.1046/j.1525 1594.1999.00751.x.

[26] Wingenfelder U., Nowak B., Furrer G., Schulin R. 2005. Adsorption of Pb and Cd by aminemodified zeolite, Water Res. 39, 3287–3297. PMID: 15996705.

[27] Perry C.C., Keeling-Tucker T., 2000. Model studies of the precipitation of silica in the presence of aluminum; implications for biology and industry. J. Inorg. Biochem., 78, 331–339. ISSN 0162-0134.

[28] Comin J.J., Barloy J., Bourrie G., Trolard F., 1999. Differential effects of monomeric and polymeric aluminium on root growth and on the biomass production of root and shoot of corn in solution culture. Eur. J. Agron., 11, 115–122. ISSN 1161-0301.

[29] Molina V.F., 2013. Soil colloids. Properties and ion binding. CRC Press, Surfactant 23 science series, vol. 156. Taylor & Francis Group, an informa business, Boca Raton-London-New York, 2013, pp. 513.

[30] Golia E.E., Dimirkou A.I., Mitsios K., 2008. Influence of some soil parameters on heavy metals accumulation by vegetables grown in agricultural soils of different soil orders. Bull. Environ. Contam. Toxicol. 81, ss. 80–84. DOI: 10.1186/s40064-015-1301-3.

[31] Davis W., 2013. Dieta bez pszenicy. Jak pozbyć się pszennego brzucha i być zdrowym (Wheat-free diet. How to get rid of the 'wheat belly' and be healthy). Wyd. Bukowy Las, 2013.

[32] Brevik E.C., Burgess L.C. (Eds.), 2012. Soils and human health. CRC Press, Taylor & 31 Francis Group, an informa business, Boca Raton-London-New York, 2012, pp. 391.